METAGRAPHS
AND THEIR APPLICATIONS

T0214578

INTEGRATED SERIES IN INFORMATION SYSTEMS

Series Editors

Professor Ramesh Sharda
Oklahoma State University

Prof. Dr. Stefan Voß
Universität Hamburg

Other published titles in the series:

E-BUSINESS MANAGEMENT: *Integration of Web Technologies with Business Models*/ edited by Michael J. Shaw

VIRTUAL CORPORATE UNIVERSITIES: *A Matrix of Knowledge and Learning for the New Digital Dawn*/ Walter R.J. Baets and Gert Van der Linden

SCALABLE ENTERPRISE SYSTEMS: *An Introduction to Recent Advances*/ edited by Vittal Prabhu, Soundar Kumara, Manjunath Kamath

LEGAL PROGRAMMING: *Legal Compliance for RFID and Software Agent Ecosystems in Retail Processes and Beyond*/ Brian Subirana and Malcolm Bain

LOGICAL DATA MODELING: *What It Is and How To Do It*/ Alan Chmura and J. Mark Heumann

DESIGNING AND EVALUATING E-MANAGEMENT DECISION TOOLS: *The Integration of Decision and Negotiation Models into Internet-Multimedia Technologies*/ Giampiero E.G. Beroggi

INFORMATION AND MANAGEMENT SYSTEMS FOR PRODUCT CUSTOMIZATION/ Thorsten Blecker et al.

MEDICAL INFORMATICS: *Knowledge Management and Data Mining in Biomedicine*/ edited by Hsinchun Chen et al.

KNOWLEDGE MANAGEMENT AND MANAGEMENT LEARNING: *Extending the Horizons of Knowledge-Based Management*/ edited by Walter Baets

INTELLIGENCE AND SECURITY INFORMATICS FOR INTERNATIONAL SECURITY: *Information Sharing and Data Mining*/ Hsinchun Chen

ENTERPRISE COLLABORATION: *On-Demand Information Exchange for Extended Enterprises*/ David Levermore and Cheng Hsu

SEMANTIC WEB AND EDUCATION/ Vladan Devedžić

INFORMATION SYSTEMS ACTION RESEARCH: *An Applied View of Emerging Concepts and Methods*/ Ned Kock

ONTOLOGIES: *A Handbook of Principles, Concepts and Applications*, edited by Raj Sharman, Rajiv Kishore and Ram Ramesh

METAGRAPHS AND THEIR APPLICATIONS

Amit Basu

Charles Wyly Professor of Information Systems
ITOM Department
Edwin L. Cox School of Business
Southern Methodist University
Dallas, TX 75275, USA

Robert W. Blanning

Professor of Management
Owen Graduate School of Management
Vanderbilt University
Nashville, TN 37203, USA

 Springer

Amit Basu
Southern Methodist University
Dallas, TX, USA

Robert W. Blanning
Vanderbilt University
Nashville, TN, USA

ISBN-13: 978-1-4419-4244-9
e-ISBN-13: 978-0-387-37234-1

Printed on acid-free paper.

9 8 7 6 5 4 3 2 1

springer.com

CONTENTS

Preface   vii

Chapter 1: Graphs, Hypergraphs, and Metagraphs   1

    1.   Graphs and Data Visualization   1

    2.   Graph Structures   4

    3.   Metagraph Theory (Part I)   9

    4.   Applications of Metagraphs (Part II)   11

Part I. Metagraph Theory   **13**

Chapter 2: The Algebraic Structure of Metagraphs   15

    1.   Formal Representation of a Metagraph   15

    2.   The Incidence and Adjacency Matrices   17

    3.   Identifying Metapaths   23

Chapter 3: Connectivity Properties of Metagraphs   27

    1.   Dominant Metapaths   27

    2.   Cutsets and Bridges   29

Chapter 4: Metagraph Transformations   33

    1.   Hierarchical Abstraction Using Projection   33

    2.   The Inverse Metagraph   46

    3.   The Element Flow Metagraph   48

Chapter 5: Attributed Metagraphs   53

    1.   Qualitative Attributes   53

    2.   Quantitative Attributes   55

    3.   Conditional Metagraphs   55

       3.1.   Projections in Conditional Metagraphs   58

       3.2.   Connectivity and Redundancy   61

Chapter 6: Independent Sub-Metagraphs   65

Part II. Applications of Metagraphs   **69**

Chapter 7: Metagraphs in Model Management   71

    1.   Models as Metagraphs   72

    2.   Model Selection and Integration   74

| 3. | Hierarchical Modeling | 76 |
| 4. | Assumptions in Model Bases | 86 |

Chapter 8: Metagraphs in Data and Rule Management 97

1.	Representing Rule Bases as Metagraphs	99
2.	Integrating Rules, Models and Data	105
3.	Discovering Implicit Integrity Constraints	111
4.	Metagraph Models of Decision Support Systems	115

Chapter 9: Metagraphs in Workflow and Process Analysis 117

1.	Representing Workflows and Processes with Metagraphs	118
2.	Views of Workflows	123
3.	Analysis of Information Interactions	128
4.	Analysis of Task Interactions	131
5.	Analysis of Resource Interactions	133
6.	Interactions among Different Types of Components	136
7.	Synthesis of Processes	137
8.	Decomposition of Processes and Implications for Organizational Design	143
9.	Representing Time-Critical Workflows with Attributed Metagraphs	146

Chapter 10: Conclusion 153

1.	The Metagraph Modeling Process	153
2.	Towards a Metagraph Workbench	156
3.	Metagraphs and Social Networks	158
4.	And Finally	160

References 161

Index of Definitions 165

Index 167

PREFACE

An important concept in the design of many information processing systems – such as transaction processing systems, decision support systems, and workflow systems – is that of a graph. In its simplest form a graph consists of a set of points (or nodes) and a set of ordered or unordered pairs of nodes (or edges). If the pairs of nodes are unordered, the graph is called a simple graph, and if they are ordered, the graph is called a directed graph, or digraph. In both cases, the graph represents a network through which materials, people, information, etc. can flow. The difference is whether the flow is restricted to one direction or whether there is no such restriction.

Simple graphs and digraphs allow for the construction of a variety of diagrammatic system design tools – such as entity-relationship diagrams, functional dependency diagrams, data flow diagrams, Petri nets, semantic nets, and the like. We note that most of these tools are representational, not analytical. That is, they provide a convenient and visually appealing format for illustrating information infrastructures, while allowing any subsequent analyses to be performed by the user.

Another problem with such graphical structures is that they usually associate individual information elements and not sets of elements. Yet in many cases it is necessary to associate sets of elements – such as multiple attributes in data relations, multiple variables in decision models, multiple logical variables in decision rules, and multiple documents in workflow systems. Furthermore, it may be necessary to integrate data relations, decision models, decision rules, and workflows into an integrated information processing system. Two multiple-element structures, hypergraphs and higraphs, allow a few such representations, but they have their limitations.

A recently developed graphical structure that overcomes the limitations and shows great promise in modeling information processing systems is a *metagraph*. Metagraphs are more complex than the graph structures described above, but they allow representation and analysis of more complex systems. Although there is a substantial literature on metagraphs, this is all in the form of journal articles and papers in conference proceedings. There have been no books presenting a comprehensive picture of the foundations of metagraphs and the applicability of these foundations to the design of information processing systems. This book attempts to fill that gap by providing a single and comprehensive treatment of metagraphs.

We begin with a brief introduction to metagraphs. A metagraph is a collection of directed set-to-set mappings. Although this is a simple definition, it leads to several powerful theoretical results and several interesting applications. We then present the material in this book in two parts. The first develops the theoretical results. Although we will include diagrams for purposes of exposition, the emphasis will be on the development of a metagraph algebra. This is a matrix algebra defined over the elements and edges of a metagraph, resulting in incidence and adjacency matrices. This in turn will lead to a more sophisticated view of paths in a metagraph, resulting in the concept of a metapath. We will also be concerned with (1) certain transformations of metagraphs, especially the projection of a metagraph to produce a simpler metagraph, (2) conditional metagraphs, in which the calculations performed early in a metagraph process determine the structure of the later part of the metagraph, and (3) submetagraphs that are largely independent of their containing metagraphs.

In the second part of the book we will examine four promising applications of metagraphs. The first is the modeling of data relations, each of which is viewed as a mapping from a set of key elements to a set of content elements. The second is the modeling of decision models, each of which is viewed as a mapping from a set of input variables to a set of output variables. The third is the modeling of decision rules, each of which is viewed as a mapping from a set of logical antecedent variables to a set of logical consequent variables. The fourth is the modeling of workflow tasks, each of which is viewed as a mapping from a set of input documents to a set of output documents. We will apply the theoretical results of the first part of the book to the application areas of the second part.

We conclude this book by briefly examining several possible extensions of this work. Of special interest is the structuring of the metagraph modeling process, which may enhance the body of work on systems analysis and design (and also software engineering), the development of a metagraph workbench to support such a process, and the possible application of our results, suitably enhanced, to social networks.

Chapter 1

GRAPHS, HYPERGRAPHS, AND METAGRAPHS

An important concept in the design of many information processing systems – such as transaction processing systems, decision support systems, project management systems, and workflow systems – is that of a graph. In its simplest form, a graph consists of a set of elements (or nodes) and a set of ordered or unordered pairs of nodes (or edges). A substantial body of theoretical and applied research on various types of graphs has made it possible to develop powerful analytical tools for systems design. The purpose of this chapter is to summarize some of the existing graph-based tools used in this area, and the purpose of this book is to present a new graphical structure, called metagraphs, that enhances existing structures and overcomes some of their disadvantages.

We begin in Section 1 by describing some of the traditional uses of graphs – tools for visualizing relationships between data elements, data aggregates, data structures, files, documents, and the like. Specifically, we examine entity-relationship diagrams, functional dependency diagrams, data flow diagrams, and semantic nets. In each of these cases the purpose of the graph is to display the structure of data so that a user can infer possible relationships of interest. Although it may be possible to use these structures as the basis of an analytical model, the purpose of the diagram/network is to assist the user's intuition in understanding important relationships among data elements, aggregates, etc.

The three remaining sections of this chapter summarize the remainder of the book. First, in Section 2 we review graph structures related to metagraphs – especially, simple graphs, directed graphs, hypergraphs, higraphs, and Petri nets. Then in Section 3 we provide a brief overview of metagraph theory, which we will examine in more detail in Chapters 2–6. Finally, in Section 4 we provide a brief overview of metagraph applications, which we will examine in more detail in Chapters 7–10. The ideas in this book are based on a set of papers published by the authors in a variety of journals and conference proceedings. These papers are included in the references at the end of the book.

1.   GRAPHS AND DATA VISUALIZATION

We begin by describing three types of graphical structures, used in three types of diagramming conventions. The first type of diagramming convention concerns the static nature of stored data – that is, the structure of databases. We will examine two approaches to diagramming databases. The first of these is based on the assumption that the data base is in relational form, so that the files (tables, relations) describe both the entities about which data is recorded, such as suppliers and the parts they supply, and the relationships among the entities. This results in the entity-relationship approach to data, illustrated in Figure 1.1.

In Figure 1.1 the supplier relation consists of two data attributes: the ID of the supplier, which is the key attribute, and the location of the supplier, which is the content attribute. Similarly, the part relation consists of two data attributes, the part ID, which is the key, and the weight of the part, which is the content. Finally, there is a many-to-many relationship between suppliers and parts, resulting in an intersection relation with a compound key (i.e., the two IDs), along with a content element (the price that the particular supplier charges for the particular part). Of course, if all suppliers charged the same price for any particular part, then the price would be a content attribute in the part relation. Thus, the structure of the data base depends on the structure of the real world about which data is being stored and/or the business rules of the organization.

Yet another approach to diagramming data bases is to focus on the functional dependencies among the data attributes. This is illustrated by the functional dependency diagram illustrated in Figure 1.2. We can see that the sup-

Figure 1.1. An entity-relationship diagram.

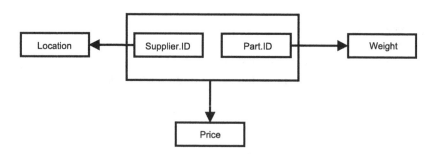

Figure 1.2. A functional dependency diagram.

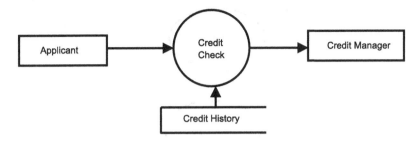

Figure 1.3. A data flow diagram.

plier ID uniquely determines location, the part ID determines weight, and the two together determine price. Both diagrams denote the same information, and both can be augmented with additional semantic information, such as the supply relationship in this case (i.e., the fact that suppliers supply parts).

The two structures outlined above describe the static structure of data in an organization, but they do not describe the dynamic nature of data as information flows throughout an organization. A common way of doing this is with a data flow diagram, illustrated in Figure 1.3. We assume that an applicant for credit submits an application, a credit check is performed, using a credit history file, and a report is sent to a credit manager. The diagram illustrates the relationships among the sources (applicant) and destinations (credit manager) of data, along with credit check process and the credit history file. This is a top-level (or Level 0) diagram, which might then be decomposed into lower-level (Levels 1, 2, etc.) diagrams, and the processes are usually numbered to make it apparent how the more detailed processes relate to each other.

Finally, we look at another type of data structure, one that describes relationships among concepts. This is captured by a semantic net, illustrated in Figure 1.4. The semantic net captures relationships among the concepts, such as instance, subclass, and others (e.g., a mouse eats cheese) and allows concepts to inherit properties from other concepts. For example, since a mouse is a mammal and Mickey Mouse is a mouse, then Mickey Mouse is a mammal. In addition, Mickey Mouse eats cheese and is an animal.

In summary, simple diagrammatic frameworks, based on graphical structures, can be used to illustrate relationships among items of interest by means of simple visualization. This allows analysts to structure the systems they must deal with and draw inferences about the behavior of these systems. But graphs can serve not only as a foundation for visualization-based inference, but they can also serve as a foundation for algebraic operations that allow for more rigorous calculation of properties.

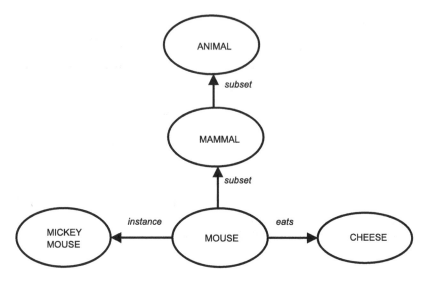

Figure 1.4. A semantic net.

2.   GRAPH STRUCTURES

We next review two traditional graph structures (ssimple graph and directed graphs), three more recent structures (hypergraphs, higraphs, and Petri nets), and finally metagraphs. To illustrate these structures, consider a system in which there are three input variables:

Pri = the sale price of a product,
Vol = the sales volume,
Wage = the prevailing wage rate.

There are also two intermediate variables:

Rev = the revenue realized, which depends on the price and the volume,
Exp = the expense incurred, which depends on the volume sold and the wage rate.

Finally there are two output variables:

Prof = the realized profit,
Notes = notes payable as a result of borrowings to cover expenses.

We assume that Pri and Vol determine Rev, Vol and Wage determine Exp, Rev and Exp determine Prof and Notes, and Exp determines Notes. We note that Notes can be determined either from Rev and Exp (along with Prof) or directly from Exp. Thus, there is a limited amount of redundancy in this set of calculation procedures, which may give the user a limited amount of discre-

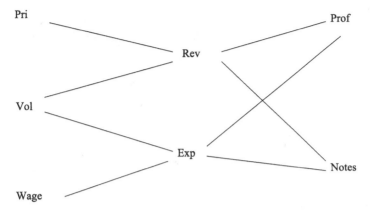

Figure 1.5. A simple graph.

tion in implementing them; however, this could lead to inconsistencies in the results.

The traditional graph structures for describing these variables and the relationships between them are simple graphs and directed graphs (Berge, 1985). A simple graph is illustrated in Figure 1.5. It consists of seven nodes, one corresponding to each of the seven variables defined above, along with seven (unordered) pairs of nodes, one for each of the edges (line segments) in the figure. Thus, we can see that there is a direct relationship between, for example, Price and Revenue, although the direction of the relationship is not clear. There is also an indirect relationship between Price and Profit; Price does not directly determine Profit, but Price does determine Profit through Revenue. The sequence of edges connecting Price to Profit is called a path. The problem is that there is also a path connecting Price to Volume, with Revenue as an intermediate node. Since we do not know the directions of the relationships, we might also conclude that Price determines Volume through Revenue, which is not the case.

A more revealing graph is a directed graph, or digraph, in which the edges are ordered pairs of nodes, represented visually by arrows. The edges of a directed graph describe the directions of the relationships among variables (nodes). This is illustrated in Figure 1.6. We can see that Price is necessary to determine Revenue, and not vice versa, and there is a path Price to Profit through Revenue. But now there is another problem. The directed graph reveals that Price and Volume determine Revenue, but it is not clear whether either Price and Volume alone are sufficient to determine Revenue, or whether both are needed. This can be overcome with AND/OR graphs, in which arcs spanning the directed edges specify whether the relationships are conjunctive or disjunctive. However, AND/OR graphs are clumsy for large numbers of

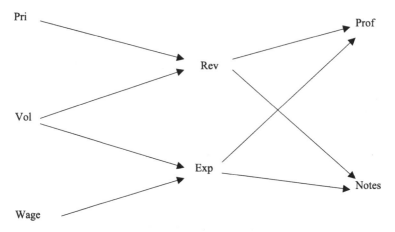

Figure 1.6. A directed graph.

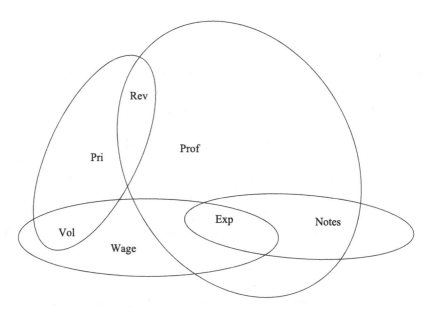

Figure 1.7. A hypergraph.

nodes and edges (i.e., variables and relationships), and a less complicated approach is needed.

A partial solution is offered by hypergraphs (Berge, 1989). In a hypergraph
each edge is a set of one or more elements, which allows us to represent relationships among multiple elements. This is illustrated in Figure 1.7. We can
see, for example, that Price, Volume, and Revenue are all part of a single relationship. As before, we can identify paths consisting of sequences of hypergraph edges connecting variables such as Price and Profit. The problem, as

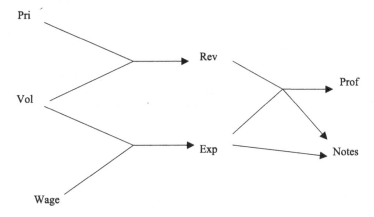

Figure 1.8. A directed hypergraph.

with simple graphs, is that the edges do not capture any sense of direction. For example, the hypergraph does not tell us whether Price and Volume are used to determine Revenue or whether some other relationship is intended – for example, that Price and Revenue are used to determine volume.

A solution is to combine the directed character of a digraph with the multivariate character of a hypergraph, resulting in a directed hypergraph, as illustrated in Figure 1.8 (Ramaswami, Sarkar and Chen, 1997). In Figure 1.8, the set {Rev, Exp} is called the tail of the edge and the set {Prof, Notes} is called the head of the edge between them. Using directed hypergraphs, we can define relationships between sets of variables, such as {Pri, Vol, Wage} and {Prof}.

Another structure is higraphs – or hierarchical graphs (Harel, 1988). A higraph is a collection of "blobs", each of which may contain elements and sub-blobs, which may in turn contain certain elements and other sub-blobs, etc. (Figure 1.9). Higraphs have the advantage of flexibility – for example, edges can originate and terminate within blobs. But this comes at the expense of analytical complexity. A related structure is statecharts (Harel, 1987), which can be used to represent sequences of calculations.

Another dynamic structure is Petri nets (Peterson, 1981). Petri nets are directed graphs containing of two types of nodes – places and transitions (Figure 1.10). Places may contain tokens, and when all of the places leading into a transition are enabled (i.e., contain at least one token), the transition may fire, removing a token from each of the places leading into it and placing a token in each place leading out of it. The process in Figure 1.10 begins with the transitions on the left side of the net firing in either order, removing the tokens from the Pri, Vol, and Wage places. A token would now appear in the Rev place and two tokens would appear in the Exp place. Now the two transitions on the right side of the net can fire, again in either order, placing tokens in Prof and Notes places. At this point no further transitions can fire and the process terminates.

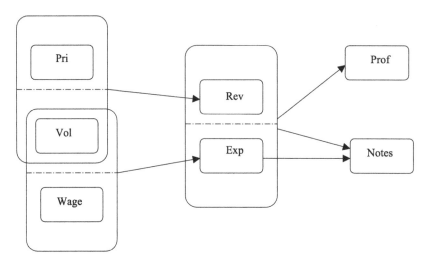

Figure 1.9. A higraph.

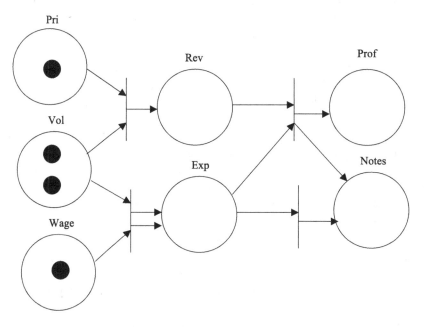

Figure 1.10. A Petri net.

Finally, we introduce the structure on which this book will focus – that of metagraphs, illustrated in Figure 1.11. A metagraph is a set of elements, which are assumed to be atomic, along with a set of edges. Each edge is an ordered pair of sets of elements, the first of which is called the invertex and the second of which is called the outvertex. Thus, metagraphs can be used to model:

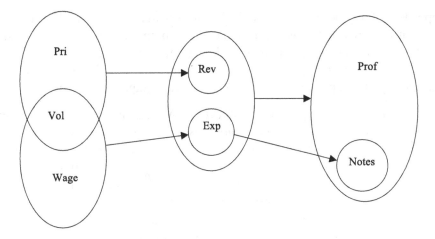

Figure 1.11. A metagraph.

- Data bases, in which invertices represent key attributes and outvertices represent content elements;
- Model bases, in which invertices represent model inputs and outvertices represent model outputs;
- Rule bases, in which invertices represent antecedent variables and outvertices represent consequent variables, and
- Workflow systems, in which invertices represent information flows entering a workstation and outvertices represent information flows emanating from a workstation.

Of these structures, the one closest to metagraphs is directed hypergraphs, in which the edges are also ordered pairs of sets of elements. The principal difference between metagraphs and directed hypergraphs is in the type of research done in these areas. Much of the work done on metagraphs is in decision support systems (DSS), and especially model-based DSS and in workflow management systems, although, as we will see, metagraphs are also relevant to other information structures, such as data management systems and rule-based systems.

We will examine two aspects of metagraphs, corresponding to the two parts of the book. The first is *Part I: Metagraph Theory*, which consists of five chapters, beginning with Chapter 2. The second is *Part II: Applications of Metagraphs*, which consists of four chapters. These are described below.

3. METAGRAPH THEORY (PART I)

Our purpose in Part I of this book is to present fundamental constructs (definitions, theorems, and interpretations of their significance) in a way that is independent of any problem context. This will provide a background for Part II in that it will provide the mathematical underpinnings essential to understanding the role that metagraphs play in analyzing the applications. But it will also provide an understanding of metagraphs that may be of assistance to anyone considering other application areas for which metagraphs may be useful.

- Chapter 2, **The Algebraic Structure of Metagraphs**, presents the use of matrices to describe metagraphs. An example is the adjacency matrix, a square matrix with one row and one column for each element in the generating set. Each member of the matrix is a set of triples, one for each edge connecting the row element to the column element. The triples define the invertex, outvertex, and the edge. We define addition and multiplication operators for the adjacency matrix, which allows us to define a transitive closure (i.e., a sum of powers) of the matrix. This will form the basis for a specification of the connectivity properties of metagraphs to be discussed in the next chapter.

- Chapter 3, **Connectivity Properties of Metagraphs**, examines the principal use of metagraphs discussed in this book. This is to determine whether there is a path connecting one set of elements to another set of elements. The definition of paths used in simple graphs and directed graphs, in which a path is a sequence of edges connecting a source element to a target element, does not apply here. Rather we define a metapath, which is a set (rather than a sequence) of edges connecting a set of source elements to a set of target elements, and this allows us to represent the parallelism found in more complex systems. In addition, we define metapath dominance, in which superfluous input elements and superfluous edges do not appear. We also investigate cycles, which are metapaths from a set of elements to itself.

- Chapter 4, **Metagraph Transformations**, examines the projection of a metagraph along a subset of its generating set. The elements in the projection consist only of those in the subset, and the edges in the projection correspond to metapaths in the original metagraph. Thus, a projection captures the connectivity relationships in a subset of a metagraph and thus represents a view of the metagraph taken by a person who is interested only in the elements contained in the projection set and the relationships between them. We also examine two related constructs, the inverse of a metagraph, in which edges become elements and elements become edges, and two related constructs – the pseudo-dual metagraph and the element-flow metagraph.

- Chapter 5, **Attributed Metagraphs**, presents an enhanced view of metagraphs in which additional variables, called attributes, are associated with the edges. One type of attribute is a resource, a qualitative or quantitative variable, which must be present (qualitative variable) or present in sufficient amount (quantitative variable) for the edge to appear in a metapath. Another type of attribute is an assumption, a Boolean variable that must be true for the edge to appear in a metapath. In this case the assumptions will be a part of the generating set – that is, the truth value of an assumption may be in the outvertex of one of the edges or may be known at the start of an analysis. A metagraph containing assumptions is called a conditional metagraph.

- Chapter 6, **Sub-metagraphs and Their Properties**, examines three concepts involving conditional metagraphs: full connectivity, non-redundancy, and independence. A set of elements B is fully connected to another set C if for every interpretation of the assumptions (i.e., every possible set of truth values for the assumptions) there is at least one metapath from B to C. B is non-redundantly connected to C if for every interpretation there is at most one metapath from B to C. A sub-metagraph within a metagraph, defined by a subset of elements and a subset of edges, is independent of the larger metagraph if the sub-metagraph interacts with the larger metagraph only through its (i.e., the sub-metagraph's) input and output elements and not through any intermediate elements. We analyze the properties of sub-metagraphs in terms of these three properties.

4. APPLICATIONS OF METAGRAPHS (PART II)

Our purpose in Part II of this book is to discuss applications of metagraphs to three areas: model management, data and rule management, and workflow and process analysis. We conclude by examining possible computational and decision support applications of metagraphs in the form of a metagraph workbench.

- Chapter 7, **Metagraphs in Model Management**, presents a model base as a metagraph, and possibly a conditional metagraph. In this representation the models are metagraph edges and the elements in the generating set are the input and output variables in the models. The connectivity properties of metagraphs can be used to determine whether a specific collection of models is sufficient to calculate a set of target variables from a set of input variables, possibly under a set of assumptions. Of special interest is hierarchical modeling, in which a composite model is composed of several base models. For example, a composite model may represent manufacturing and marketing relationships extracted from two base models. In metagraph terms, the composite model is represented by combining projections of the base models. We examine the relationship between the sum of two or more projections and the projection of the sum.

- Chapter 8, **Metagraphs in Data and Rule Management**, extends the analyses of the previous chapter to encompass metagraph edges as functional dependencies in data bases and as Horn clauses in rule bases. The elements in the generating set are data attributes (in the case of data bases) and propositions (in the case of rule bases). Metapaths represent inference paths from a set of source variables to a set of target variables under most conditions. The requirement is that there be an acyclic metapath from the source elements to the target elements. In addition, metagraphs can be used to uncover implicit integrity constraints in rule bases. Thus, metagraphs can be used to integrate data bases, decision models, and rules in a decision support system.
- Chapter 9, **Metagraphs in Workflow and Process Analysis**, examines the use of metagraphs in modeling workflow support systems. In this case the elements in the generating represent are information elements, often in the form of paper or electronic documents, and the edges represent workstations at which document processing takes place (e.g., extraction of credit information from a loan application). We will be concerned with the decomposition of workflows (e.g., to identify candidates for outsourcing) and the synthesis of separate workflows (e.g., to consolidate interdependent processes). We will also offer comments about the impact of these considerations on organizational design.
- Chapter 10, **Conclusion**, addresses three issues. The first is the metagraph modeling process and specifically the life cycle of metagraph construction and implementation. The second is the concept of a metagraph workbench that will assist a modeler in constructing and implementing metagraphs. The third is a discussion of yet another promising metagraph application area – the use of metagraphs in modeling social networks.

PART I

METAGRAPH THEORY

Chapter 2

THE ALGEBRAIC STRUCTURE OF METAGRAPHS

In Chapter 1, the notion of a metagraph was introduced informally, using visual depictions and descriptions. In this chapter, the formal structure of a metagraph is defined, and its basic properties are identified.

1. FORMAL REPRESENTATION OF A METAGRAPH

DEFINITION 2.1. The *generating set* of a metagraph is the set of *elements* $X = \{x_1, x_2, \ldots, x_n\}$, which represent variables of interest, and which occur in the edges of the metagraph.

DEFINITION 2.2. An *edge e* in a metagraph is a pair $e = \langle V_e, W_e \rangle \in E$ (where E is the set of edges) consisting of an *invertex* $V_e \subset X$ and an *outvertex* $W_e \subset X$, each of which may contain any number of elements. The different elements in the invertex (outvertex) are *coinputs* (*cooutputs*) of each other.

DEFINITION 2.3. A *metagraph* $S = \langle X, E \rangle$ is then a graphical construct specified by its generating set X and a set of edges E defined on the generating set.

DEFINITION 2.4. A *simple path* $h(x, y)$ from an element x to an element y is a sequence of edges $\langle e_1, e_2, \ldots, e_n \rangle$ such that

$x \in invertex(e_1)$,
$y \in outvertex(e_n)$, and
for all $e_i, i = 1, \ldots, n - 1, outvertex(e_i) \cap invertex(e_{i+1}) \neq \varnothing$.

The *coinput* of x in the path (denoted $coinput(x)$) is the set of all other invertex elements in the path's edges that are not also in the outvertex of any edges in the path, and the *cooutput* of y (denoted $cooutput(y)$) is the set of all outvertex elements other than y. The *length* of a simple path is the number of edges in the path.

EXAMPLE 2.1. The metagraph in Figure 2.1 can be represented as follows:

$S = \langle X, E \rangle$, where
$X = \{Exp, Notes, Prof, Rev, Pri, Vol, Wage\}$, and

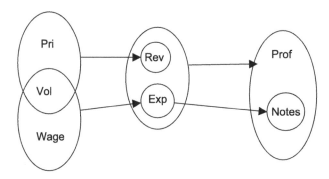

Figure 2.1. An example metagraph.

$E = \{\langle\{Pri, Vol\}, \{Rev\}\rangle, \langle\{Vol, Wage\}, \{Exp\}\rangle, \langle\{Rev, Exp\},$
$\quad\quad \{Prof, Notes\}\rangle, \langle\{Exp\}, \{Notes\}\rangle\},$
$Invertex(\langle\{Rev, Exp\}, \{Prof, Notes\}\rangle) = \{Rev, Exp\},$
$Outvertex(\langle\{Rev, Exp\}, \{Prof, Notes\}\rangle) = \{Prof, Notes\},$
$Coinput(Rev, \langle\{Rev, Exp\}, \{Prof, Notes\}\rangle) = \{Exp\},$
$Cooutput(Prof, \langle\{Rev, Exp\}, \{Prof, Notes\}\rangle) = \{Notes\}.$

The edges of S can be labeled, so that for example, $e_1 = \langle\{Rev, Exp\},$
$\{Prof, Notes\}\rangle$.

Note that a single metagraph edge is a singular metagraph. Also, note that
an edge with a singular invertex and a singular outvertex is isomorphic with
an edge in a directed graph.

Simple paths do not describe all of the connectivity properties of metagraphs. This is illustrated in the metagraph of Figure 2.2, in which there are
two simple paths from x_1 to x_5, both of which have non-null coinputs. However, x_1 itself is sufficient to calculate x_5, if all three edges e_1, e_2, and e_3 are
used. However, $\langle e_1, e_2, e_3 \rangle$ does not represent a simple path, since there is no
sequence of connected edges consisting of these edges. Rather, this metapath
is the union of edges in two simple paths.

DEFINITION 2.5. Given a metagraph $S = \langle X, E \rangle$, a *metapath $M(B, C)$* from
a source $B \subset X$ to a target $C \subset X$ is a set of edges $E' \subseteq E$ such that (1) each
$e' \in E'$ is on a simple path from some element in B to some element in C,
(2) $[\bigcup_{e'} V_{e'} \setminus \bigcup_{e'} W_{e'}] \subseteq B$, and (3) $C \subseteq \bigcup_{e'} W_{e'}$.

There are three differences between simple paths and metapaths:

- First, a metapath is a set of edges and not a sequence of edges. For example, in Figure 2.2, one metapath from x_1 to x_5 is $M(\{x_1\}, \{x_5\}) =$
  $\{e_1, e_2, e_3\}$.

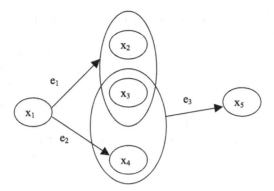

Figure 2.2. Metapath example.

- Second, the source and target of a metapath are sets, not elements, as in simple paths. Of course, these sets may sometimes be singleton sets, as is the case in Figure 2.2 (with $B = \{x_1\}$ and $C = \{x_5\}$).
- Third, the notion of a coinput does not apply to a metapath, since the source set includes all pure inputs.

2.   THE INCIDENCE AND ADJACENCY MATRICES

In order to define an algebra for metagraph manipulation, two matrix representations of a metagraph are needed. These are the adjacency matrix and incidence matrix, respectively. It is worth noting that as with traditional graph structures, each of these matrices is a complete representation of a metagraph, and can be derived from the other.

DEFINITION 2.6. The *adjacency matrix* A for a metagraph $S = \langle X, E \rangle$ is an $I \times I$ matrix (where $I = |X|$), such that for all $i, j \in \{1, \ldots, I\}$,

$$a_{ij} = \bigcup_k (\alpha_{ij})_k,$$

where

$$(\alpha_{ij})_k = \begin{cases} \langle V_k \backslash \{x_i\}, W_k \backslash \{x\}, \langle E_k \rangle \rangle & \text{if } x_i \in V_k \wedge x_j \in W_k, \\ \phi & \text{otherwise.} \end{cases}$$

In other words, the adjacency matrix A of a metagraph is a square matrix with one row and one column for each element in the generating set X. The ijth element of A, denoted a_{ij}, is a set of triples, one for each edge e connecting x_i to x_j. Each triple is of the form $\langle CI_e, CO_e, e \rangle$, in which CI_e is the coinput of x_i in e and CO_e is the cooutput of x_j in e.

For example, the adjacency matrix for the metagraph in Figure 2.3 below is shown in Figure 2.4.

There is an algebra defined for metagraph adjacency matrices. Given adjacency matrices A_1 and A_2, defined for two metagraphs that have the same generating set, these matrices can be added and multiplied with the result in each case being another matrix over the same generating set. Intuitively, $A_1 + A_2$ represents the adjacency matrix of the union of the two metagraphs, while $A_1 * A_2$ represents all paths of length two, where the first edge is from the first metagraph and the second edge is from the second metagraph.

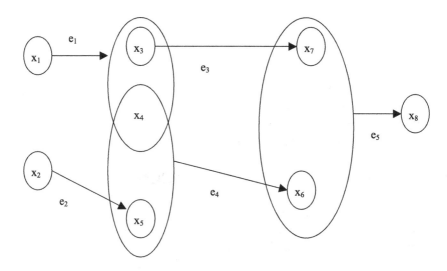

Figure 2.3. An example metagraph.

	x_1	x_2	x_3	x_4	x_5	x_6	x_7	x_8
x_1	$\varnothing$	$\varnothing$	$\langle\phi,\{x_4\},e_1\rangle$	$\langle\phi,\{x_3\},e_1\rangle$	$\varnothing$	$\varnothing$	$\varnothing$	$\varnothing$
x_2	$\varnothing$	$\varnothing$	$\varnothing$	$\varnothing$	$\langle\phi,\phi,e_2\rangle$	$\varnothing$	$\varnothing$	$\varnothing$
x_3	$\varnothing$	$\varnothing$	$\varnothing$	$\varnothing$	$\varnothing$	$\varnothing$	$\langle\phi,\phi,e_3\rangle$	$\varnothing$
x_4	$\varnothing$	$\varnothing$	$\varnothing$	$\varnothing$	$\varnothing$	$\langle\{x_5\},\phi,e_4\rangle$	$\varnothing$	$\varnothing$
x_5	$\varnothing$	$\varnothing$	$\varnothing$	$\varnothing$	$\varnothing$	$\langle\{x_4\},\phi,e_4\rangle$	$\varnothing$	$\varnothing$
x_6	$\varnothing$	$\varnothing$	$\varnothing$	$\varnothing$	$\varnothing$	$\varnothing$	$\varnothing$	$\langle\{x_7\},\phi,e_5\rangle$
x_7	$\varnothing$	$\varnothing$	$\varnothing$	$\varnothing$	$\varnothing$	$\varnothing$	$\varnothing$	$\langle\{x_6\},\phi,e_5\rangle$
x_8	$\varnothing$	$\varnothing$	$\varnothing$	$\varnothing$	$\varnothing$	$\varnothing$	$\varnothing$	$\varnothing$

Figure 2.4. The adjacency matrix for the metagraph in Figure 2.3.

	x_1	x_2	x_3	x_4	x_5	x_6	x_7	x_8
x_1	$\varnothing$	$\varnothing$	$\varnothing$	$\varnothing$	$\varnothing$	$\varnothing$	$\varnothing$	$\varnothing$
x_2	$\varnothing$	$\varnothing$	$\varnothing$	$\varnothing$	$\varnothing$	$\varnothing$	$\varnothing$	$\varnothing$
x_3	$\varnothing$	$\varnothing$	$\varnothing$	$\varnothing$	$\varnothing$	$\langle\{x_4\},\phi,e_6\rangle$	$\varnothing$	$\varnothing$
x_4	$\langle\phi,\phi,e_7\rangle$	$\varnothing$	$\varnothing$	$\varnothing$	$\varnothing$	$\langle\{x_3\},\phi,e_6\rangle$	$\varnothing$	$\varnothing$
x_5	$\varnothing$	$\varnothing$	$\varnothing$	$\varnothing$	$\varnothing$	$\varnothing$	$\varnothing$	$\varnothing$
x_6	$\varnothing$	$\varnothing$	$\varnothing$	$\varnothing$	$\varnothing$	$\varnothing$	$\varnothing$	$\varnothing$
x_7	$\varnothing$	$\varnothing$	$\varnothing$	$\varnothing$	$\varnothing$	$\varnothing$	$\varnothing$	$\varnothing$
x_8	$\varnothing$	$\varnothing$	$\varnothing$	$\varnothing$	$\varnothing$	$\varnothing$	$\varnothing$	$\varnothing$

Figure 2.5. Adjacency matrix of additional metagraph.

DEFINITION 2.7. Given a generating set X and two metagraphs $S_1 = \langle X, E_1 \rangle$ and $S_2 = \langle X, E_2 \rangle$ with adjacency matrices A_1 and A_2 respectively, then the *sum* of the two adjacency matrices is the adjacency matrix of the metagraph $S_3 = \langle X, E_1 \cup E_2 \rangle$ with components

$$(A_1 + A_2)_{ij} = a_{ij}^1 \cup a_{ij}^2.$$

Note that the two matrices must be defined on the same generating set. However, this is not a restrictive requirement. If the generating sets of the two metagraphs are overlapping but not identical, each metagraph can be defined over a new generating set which is the union of the two generating sets, and then the above definition can be applied.

As an example, consider the metagraph in Figure 2.3 combined with a metagraph consisting of two edges, $e_6 = \langle \{x_3, x_4\}, \{x_6\} \rangle$ and $e_7 = \langle \{x_4\}, \{x_1\} \rangle$, which has the adjacency matrix shown in Figure 2.5.

The result of adding the two adjacency matrices gives the adjacency matrix of the union of the two metagraphs, and this is shown in Figure 2.6.

The definition of multiplication of adjacency matrices is computationally more complex, since the result is not an adjacency matrix, but rather a matrix that identifies paths of length two between elements, as mentioned above. In order to define this operator, a number of preliminary concepts need to be specified.

DEFINITION 2.8. The *components* of an ordered triple R are $\alpha(R)$, $\beta(R)$ and $\gamma(R)$ respectively (i.e., $R = \langle \alpha(R), \beta(R), \gamma(R) \rangle$).

	x_1	x_2	x_3	x_4	x_5	x_6	x_7	x_8
x_1	$\varnothing$	$\varnothing$	$\langle\phi,\{x_4\},e_1\rangle$	$\langle\phi,\{x_3\},e_1\rangle$	$\varnothing$	$\varnothing$	$\varnothing$	$\varnothing$
x_2	$\varnothing$	$\varnothing$	$\varnothing$	$\varnothing$	$\langle\phi,\phi,e_2\rangle$	$\varnothing$	$\varnothing$	$\varnothing$
x_3	$\varnothing$	$\varnothing$	$\varnothing$	$\varnothing$	$\varnothing$	$\langle\{x_4\},\phi,e_6\rangle$	$\langle\phi,\phi,e_3\rangle$	$\varnothing$
x_4	$\langle\phi,\phi,e_7\rangle$	$\varnothing$	$\varnothing$	$\varnothing$	$\varnothing$	$\langle\{x_5\},\phi,e_4\rangle,$ $\langle\{x_3\},\phi,e_6\rangle$	$\varnothing$	$\varnothing$
x_5	$\varnothing$	$\varnothing$	$\varnothing$	$\varnothing$	$\varnothing$	$\langle\{x_4\},\phi,e_4\rangle$	$\varnothing$	$\varnothing$
x_6	$\varnothing$	$\varnothing$	$\varnothing$	$\varnothing$	$\varnothing$	$\varnothing$	$\varnothing$	$\langle\{x_7\},\phi,e_5\rangle$
x_7	$\varnothing$	$\varnothing$	$\varnothing$	$\varnothing$	$\varnothing$	$\varnothing$	$\varnothing$	$\langle\{x_6\},\phi,e_5\rangle$
x_8	$\varnothing$	$\varnothing$	$\varnothing$	$\varnothing$	$\varnothing$	$\varnothing$	$\varnothing$	$\varnothing$

Figure 2.6. The adjacency matrix for the combined metagraph.

DEFINITION 2.9. The operator $Cat(A, B)$ represents the concatenation of two ordered lists A and B.

For example, $Cat(\langle q, r\rangle, \langle q, s, t\rangle) = \langle q, r, q, s, t\rangle$.

DEFINITION 2.10. The $Trnc(.)$ operator truncates a list when it encounters a duplicate element.

For example, $Trnc(\langle a_n, \ n = 1, \ldots, N\rangle) = \langle a_n, n = 1, \ldots, M\rangle$, where $Q = \{a_n, \ n = 1, \ldots, M\}$ is a set of distinct elements and $a_{M+1} \in Q$.

DEFINITION 2.11. Let X be a generating set and let two metagraphs with adjacency matrices A and B respectively be defined on this generating set. Each cell in these matrices is a list of triples, with the nth triple in a_{ik} and the mth triple in b_{kj} denoted as $(a_{ik})_n$ and $(b_{kj})_m$ respectively. Then the 'o' operator defines either an ordered triple or a null set, as follows:

(1) If $((a_{ik})_n \neq \phi) \wedge ((b_{kj})_m \neq \phi)$ then $(a_{ik})_n \circ (b_{kj})_m$ is a triple R specified as follows:
  (a) $\alpha(R) = (\alpha((a_{ik})_n) \cup \alpha((b_{kj})_m)) \backslash (\beta((a_{ik})_n) \cup \{x_i\})$,
  (b) $\beta(R) = (\beta((a_{ik})_n) \cup \beta((b_{kj})_m) \cup \{x_k\}) \backslash \{x_j\}$,

(c) $\gamma(R) = Trnc(Cat(\gamma(a_{ik})_n, \gamma(b_{kj})_m))$;

(2) Else $(a_{ik})_n \circ (b_{kj})_m = \phi$.

DEFINITION 2.12. Given a generating set X and two metagraphs $S_1 = \langle X, E_1 \rangle$ and $S_2 = \langle X, E_2 \rangle$ with adjacency matrices A and B respectively, let $(a_{ij})_n$ and $(b_{ij})_n$ be the ordered triples in a_{ij} and b_{ij} such that:

$$a_{ik} = \{(a_{ik})_n, \ n = 1, \ldots, N\} \quad \text{and} \quad b_{kj} = \{(b_{kj})_m, \ m = 1, \ldots, M\}.$$

Then the *product* of the two adjacency matrices A and B is denoted $A \times B$ with components

$$(A \times B)_{ij} = \bigcup_{k=1}^{K} \bigcup_{n=1}^{N} \bigcup_{m=1}^{M} \left((a_{ik})_n \circ (b_{kj})_m\right).$$

EXAMPLE 2.2. Given $a_{ik} = \langle \phi, \{x_4\}, e_1 \rangle$ and $b_{kj} = \{\langle \{x_2\}, \{x_5\}, e_2 \rangle, \langle \{x_4\}, \phi, e_3 \rangle\}$, consider the first combination of $(a_{ik})_1 \circ (b_{kj})_1$. Since neither of them is null, we get a triple as follows:

$$\alpha\left((a_{ik})_1 \circ (b_{kj})_1\right) = \left(\phi \cup \{x_2\}\right) \backslash \left(\{x_4\} \cup \{x_1\}\right) = \{x_2\},$$

$$\beta\left((a_{ik})_1 \circ (b_{kj})_1\right) = \left(\{x_4\} \cup \{x_5\} \cup \{x_3\}\right) \backslash \{x_6\} = \{x_3, x_4, x_5\},$$

$$\gamma\left((a_{ik})_1 \circ (b_{kj})_1\right) = Trnc\left(Cat(\langle e_1 \rangle, \langle e_2 \rangle)\right) = \langle e_1, e_2 \rangle.$$

Similarly, $(a_{ik})_1 \circ (b_{kj})_2 = \langle \phi, \{x_3, x_4\}, \langle e_1, e_3 \rangle \rangle$.

Using multiplication, the powers of an adjacency matrix can also be computed. The nth power of A is denoted A^n. The ijth element of A^n, denoted a_{ij}^n, is a set of triples, one for each simple path $h(x_i, x_j)$ of length n connecting x_i to x_j. Each triple is of the form $\langle CI_h, CO_h, h \rangle$, in which h denotes the sequence of edges comprising the path, CI_h is the coinput of x_i in h and CO_h is the cooutput of x_j in h. The *closure* of A, denoted $A^* = A + A^2 + \cdots$, represents all simple paths of any length in the metagraph. The ijth element of A^*, denoted a_{ij}^*, is a set of triples, one for each simple path $h(x_i, x_j)$ of any length connecting x_i to x_j. Note that the multiplication operator allows any cycle to be traversed only once. Figure 2.7 shows the closure of the adjacency matrix in Figure 2.4.

The addition and multiplication operators on adjacency matrices of metagraphs also support the properties of associativity and distributivity, as shown below:

	x_1	x_2	x_3	x_4	x_5	x_6	x_7	x_8
x_1	$\varnothing$	$\varnothing$	$\langle\phi,\{x_4\},e_1\rangle$	$\langle\phi,\{x_3\},e_1\rangle$	$\varnothing$	$\langle\{x_3\},\{x_3,x_4\},\langle e_1,e_3\rangle\rangle$	$\langle\phi,\{x_3,x_4\},\langle e_1,e_3\rangle\rangle$	$\{\langle\{x_3\},\{x_3,x_4,x_7\},\langle e_1,e_3,e_5\rangle\rangle,$ $\langle\{x_5,x_7\},\{x_3,x_4,x_6\},\langle e_1,e_4,e_5\rangle\rangle\}$
x_2	$\varnothing$	$\varnothing$	$\varnothing$	$\varnothing$	$\langle\phi,\phi,e_2\rangle$	$\langle\{x_4\},\{x_3\},\langle e_2,e_4\rangle\rangle$	$\varnothing$	$\{\langle\{x_4,x_7\},\{x_3,x_6\},\langle e_2,e_4,e_5\rangle\rangle\}$
x_3	$\varnothing$	$\varnothing$	$\varnothing$	$\varnothing$	$\varnothing$	$\varnothing$	$\langle\phi,\phi,e_2\rangle$	$\{\langle\{x_6\},\{x_7\},\langle e_3,e_5\rangle\rangle\}$
x_4	$\varnothing$	$\varnothing$	$\varnothing$	$\varnothing$	$\varnothing$	$\langle\{x_5\},\phi,e_4\rangle$	$\varnothing$	$\{\langle\{x_5,x_7\},\{x_6\},\langle e_4,e_5\rangle\rangle\}$
x_5	$\varnothing$	$\varnothing$	$\varnothing$	$\varnothing$	$\varnothing$	$\langle\{x_4\},\phi,e_4\rangle$	$\varnothing$	$\{\langle\{x_4,x_7\},\{x_6\},\langle e_4,e_5\rangle\rangle\}$
x_6	$\varnothing$	$\varnothing$	$\varnothing$	$\varnothing$	$\varnothing$	$\varnothing$	$\varnothing$	$\langle\{x_7\},\phi,e_5\rangle$
x_7	$\varnothing$	$\varnothing$	$\varnothing$	$\varnothing$	$\varnothing$	$\varnothing$	$\varnothing$	$\langle\{x_6\},\phi,e_5\rangle$
x_8	$\varnothing$	$\varnothing$	$\varnothing$	$\varnothing$	$\varnothing$	$\varnothing$	$\varnothing$	$\varnothing$

Figure 2.7. The closure of the adjacency matrix in Figure 2.4.

THEOREM 2.1. *Given a generating set X and three metagraphs defined on this set with adjacency matrices A, B, and C respectively, then*

(1) $A \times (B \times C) = (A \times B) \times C$,

(2) $A + (B \times C) = (A \times C) + (B \times C)$.

PROOF. Since the multiplication operation identifies all paths made up of edges in the first metagraph followed by an edge in the second metagraph, $A \times (B \times C)$ identifies all paths of length three consisting of an edge from A followed by an edge from B and then an edge from C respectively. This is the same as in $(A \times B) \times C$, which proves associativity.

To prove the distributive property, if $D = (A \times C) + (B \times C)$, it suffices to show that for any $i, j, d_{ij} = ((A + B) \times C)_{ij}$. In the following, the notation $(a_{ij})_n$ refers to the nth triple in a_{ij}, while a_{ij}^n refers to the entry in the ith row and jth column of A^n, and $(a_{ij}^n)_m$ refers to the mth element of a_{ij}^n.

Let $|a_{ij}| = M_1$, $|b_{ij}| = M_2$, and $|c_{ij}| = N$. Also, let $Y = A + B$ (i.e., $\forall i, j, y_{ij} = a_{ij} \cup b_{ij}$. Reorganize b_{ij} so that for $q \leq Q$, $(b_{ij})_q \notin a_{ij}$ and for all $q > Q$, $(b_{ij})_q \in a_{ij}$. Thus, $|y_{ij}| = M_1 + Q$. Then x_{ij} can be partitioned into the following sets:

$$(y_{ij})_p = (a_{ij})_{m_1} \quad \text{for } p = 1, \ldots, M_1,$$

$$(y_{ij})_p = (b_{ij})_{p-M_1} \quad \text{for } p = (M_1 + 1), \ldots, (M_1 + Q).$$

Then,

$$d_{ij} = \left(\bigcup_{k,m_1,n} (a_{ik})_{m_1} \circ (c_{kj})_n \right) \cup \left(\bigcup_{k,m_2,n} (b_{ik})_{m_2} \circ (c_{kj})_n \right)$$

$$= \bigcup_{k,n} \left(\left(\bigcup_{m_1} (a_{ik})_{m_1} \circ (c_{kj})_n \right) \cup \left(\bigcup_{m_2} (b_{ik})_{m_2} \circ (c_{kj})_n \right) \right)$$

$$= \bigcup_{k,n} \left(\left(\bigcup_{p=1}^{M_1} (y_{ik})_p \circ (c_{kj})_n \right) \cup \left(\bigcup_{p=M_1+1}^{M_1+Q} (y_{ik})_p \circ (c_{kj})_n \right) \right)$$

$$= \bigcup_{k,n} \left(\bigcup_{p=1}^{M_1+Q} (y_{ik})_p \circ (c_{kj})_n \right)$$

$$= \bigcup_{k,p,n} \left((a_{ik} \cup b_{ik})_p \circ (c_{kj})_n \right).$$

Thus, $D = (A + B) \times C$, which is the desired result. $\qquad\square$

Also, note that the null matrix D (with $d_{ij} = \phi \; \forall i, j$) is a left and right identity under addition (i.e., $A + D = D + A = A$). This implies that the set of all adjacency matrices defined on the same generating set forms a commutative idempotent monoid under addition, while the set of all non-null adjacency matrices forms a semi-group under multiplication.

DEFINITION 2.13. The *incidence matrix* G of a metagraph has one row for each element in the generating set and one column for each edge. The ijth component of G, g_{ij}, is -1 if x_i is in the invertex of e_j, it is $+1$ if x_i is in the outvertex of e_j, and it is $\emptyset$ otherwise.

The incidence matrix for the metagraph in Figure 2.3 is shown in Figure 2.8 below.

Once the closure A^* of a metagraph's adjacency matrix has been constructed, it can be used to identify a variety of connectivity features of that metagraph, as discussed in the next chapter.

3. IDENTIFYING METAPATHS

The adjacency matrix and its closure can be used to find paths and metapaths. One of the benefits of the metagraph representation (versus simpler graph representations) is that searches for metapaths can be limited to only

	e_1	e_2	e_3	e_4	e_5
x_1	-1				
x_2		-1			
x_3	+1		-1		
x_4	+1			-1	
x_5		+1		-1	
x_6				+1	-1
x_7			+1		-1
x_8					+1

Figure 2.8. The incidence matrix for the metagraph in Figure 2.3.

those portions of the A^* matrix that deals with the elements in the source and target sets, which can substantially reduce the search space. This is because every metapath from B to C must consist of edges based on a combination of triples from cells a_{ij}^* such that $x_i \in B$ and $x_j \in C$. Furthermore, the efficiency of the search procedure now becomes a function of the number of simple paths between B and C (each corresponding to a triple in the candidate set), rather than the entire closure matrix.

Another useful observation that can be exploited is that if there is a metapath from B to C, then there should be triples composed of these edges in A^* in every column j such that $x_j \in C$. Also, in using the closure matrix to find metapaths $M(B, C)$, even though there is at least one triple in every column of A^* corresponding to elements of C, it is not always necessary to examine each triple explicitly, because the triples include the co-inputs and co-outputs for the path that they represent. Furthermore, if we use a conservative approach that always considers a minimal number of rows, then the metapaths obtained are all input-dominant.

DEFINITION 2.14. Given a metagraph $S = \langle X, E \rangle$, for any two sets of elements B and C in S, a metapath $M(B, C)$ is said to be *input-dominant* if there is no metapath $M'(B', C)$ such that $B' \subset B$.

Based on the above observations, the procedure to find metapaths can be described as follows:

1. Select a candidate set of input rows I in A^* such that $x_i \in B, \forall i \in I, B = \bigcup_i x_i$. Start with single rows, and repeat with larger sets progressively in successive iterations.
2. If $\exists x_j \in B$ such that $a^*_{ij} = \phi, \forall i \in I$, then there is no metapath from $\{x_i \mid i \in I\}$ to C. Return to step 1 and repeat with another set of rows.
3. Find a candidate set of triples in cells a^*_{ij} such that $i \in I, x_j \in C$ that forms a cover for C (where a cover for C is a set of triples T such that $C \subseteq \bigcup_{t \in T} output(t)$). If such a cover is found, then $\bigcup_{t \in T} path(t)$ comprises an input dominant metapath from $B(= \{x_i \mid i \in I\})$ to C.
4. Otherwise, return to step 1 and use an alternative candidate set I.

The stopping criterion for the procedure (in step 3 after a metapath is found, or in step 1 if there are no more candidate sets) depends upon whether the desired outcome is one metapath or every metapath.

Chapter 3

CONNECTIVITY PROPERTIES OF METAGRAPHS

In this chapter, we further develop the connectivity features of paths and metapaths introduced in Chapter 2. In particular, we introduce the notions of bridges, cycles and the properties of dominance.

1.   DOMINANT METAPATHS

The property of dominance is useful in determining whether a metapath has any unnecessary components (edges or elements). We introduce the concept constructively, based on the following definitions:

DEFINITION 3.1. Given a metagraph $S = \langle X, E \rangle$, for any two sets of elements B and C in S, a metapath $M(B, C)$ is said to be *edge-dominant*   if no proper subset of $M(B, C)$ is also a metapath from B to C.

DEFINITION 3.2 (also Definition 2.14). Given a metagraph $S = \langle X, E \rangle$, for any two sets of elements B and C in S, a metapath $M(B, C)$ is said to be *input-dominant* if there is no metapath $M'(B', C)$ such that $B' \subset B$.

In other words, edge-dominance (input-dominance) ensures that none of the edges (elements) in the metapath is superfluous or dispensable. Based on these concepts, we can then define a dominant metapath as follows:

DEFINITION 3.3. Given a metagraph $S = \langle X, E \rangle$, for any two sets of elements B and C in S, a metapath $M(B, C)$ is said to be *dominant* if it is both edge-dominant and input-dominant.

We can illustrate these concepts using an example. Consider the metagraph shown in Figure 3.1, which consists of seven edges interconnecting nine elements.

In this metagraph, $M_1(\{x_1, x_4\}, \{x_6\}) = \{e_1, e_2, e_3\}$ is an edge-dominant metapath from $\{x_1, x_4\}$ to $\{x_6\}$. It is also an input-dominant metapath, since there is no metapath from any proper subset of $\{x_1, x_4\}$ to $\{x_6\}$. Thus, by definition it is a dominant metapath for that given source–target pair of sets as well.

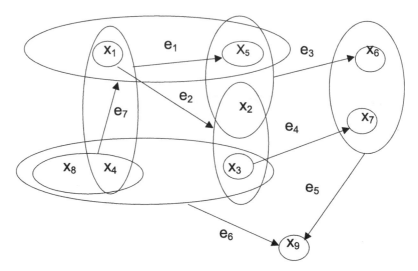

Figure 3.1. Example metagraph.

Consider now the metapath $M_2(\{x_3, x_4, x_8\}, \{x_9\}) = \{e_6\}$. Since it is a single edge, this metapath is by definition an edge-dominant metapath (since it does not have any non-empty proper subset of edges). However, it is interesting to note that it is not an input-dominant metapath from its invertex to its outvertex. This is because there exists a metapath $M_3(\{x_4, x_8\}, \{x_9\}) = \{e_7, e_2, e_3, e_4, e_5\}$. Even though this metapath involves several edges, its pure inputs are a proper subset of the invertex of e_6. This is interesting, because it illustrates that even a single edge could be dominated by some other set of edges, and in this sense, metagraphs are distinctly different from simpler graph structures.

The algebraic representation of a metagraph in terms of its adjacency matrix A can be used to identify dominant metapaths. In Chapter 2, we discussed how all the metapaths between two sets B and C can be found from the A and A^* matrices.

DEFINITION 3.4. An edge e is said to be *non-redundant* in a metapath $M(B, C)$ if there is some $Y \subset C$ such that for every metapath M' from B to Y, $e \in M'(B, Y)$.

Once again, in Figure 3.1, we see that e_1 is non-redundant in the metapath $M_1(\{x_1, x_4\}, \{x_5, x_6\}) = \{e_1, e_2, e_3\}$, since it is in every metapath from the set $\{x_1, x_4\}$ to $\{x_5\}$. At the same time, the same edge e_1 is redundant in the metapath $M_4(\{x_4, x_8\}, \{x_5, x_6\}) = \{e_1, e_2, e_3, e_4, e_7\}$, since the subset $\{x_5\}$ is now reachable by e_7 already, so that e_1 is no longer needed.

THEOREM 3.1. *A metapath $M(B, C)$ is edge-dominant iff each of its edges is non-redundant in $M(B, C)$.*

PROOF. Since every edge e in the metapath is non-redundant, removal of e will disconnect at least one of the elements in C from B. Thus, every edge in M is essential for the metapath, making the metapath edge-dominant. $\square$

In Chapter 2, we showed how the algebraic representation of a metagraph's structure can be used to find metapaths. Testing a candidate metapath for dominance is straightforward. Since the procedure implicitly generates input-dominant metapaths, only edge-dominance needs to be tested. By eliminating each edge in turn from the metapath and testing the resulting target set against C, the edge-dominance of $M(B, C)$ can be tested in time proportional to $|M(B, C)|$.

2. CUTSETS AND BRIDGES

The notion of a bridge addresses the extent of connectivity between sets of elements in a metagraph. It is defined using a related concept, a cutset.

DEFINITION 3.5. Given two sets of elements B and C in a metagraph $S = \langle X, E \rangle$, such that there is a metapath $M(B, C)$, a set of edges E' is a *cutset* between B and C if there is no metapath from B to C in $S' = \langle X, E \backslash E' \rangle$; furthermore, there is no proper subset of E' that is also a cutset between B and C.

DEFINITION 3.6. A singleton cutset between two element sets B and C is a *bridge* between them.

These concepts are useful in many applications, since they allow designers and analysts to focus on edges that are critical. Note that a cutset does not just affect a particular metapath $M(B, C)$ between the source and target sets B and C; it removes *every* metapath between them. It should also be easy to see that there could be multiple cutsets between any pair of sets. Intuitively, any set of edges made up of one edge from every edge dominant metapath between B and C comprises a cutset between B and C. Thus, if there are two edge-disjoint metapaths $M_1(B, C)$ and $M_2(B, C)$ between B and C, then the number of cutsets between B and C is $|M_1| \cdot |M_2|$.

On the other hand, even though there may be many cutsets between two sets B and C in a metagraph, there may be still no bridges between them. For instance, in the above example of two edge-disjoint metapaths between B and C, there is clearly no bridge between B and C.

We can also illustrate these concepts using our example metagraph in Figure 3.1. In that figure, we see that $\{x_4, x_8\}$ is connected to both $\{x_7\}$ and $\{x_9\}$. It turns out that there is no bridge between $\{x_4, x_8\}$ and $\{x_9\}$, and there is at least one metapath between them even if any single edge in the metagraph is removed. On the other hand, each of the edges e_2, e_4, and e_7 is a bridge between $\{x_4, x_8\}$ and $\{x_7\}$, since the removal of any of these edges disconnects the source from the target.

The following theorems provide some useful properties of bridges in metagraphs:

THEOREM 3.2. *Let B and C be two disjoint sets of elements in a metagraph, such that there is at least one metapath from B to C. If there is no bridge between B and C, then there are at least two edge-dominant metapaths from B to C.*

PROOF. Since C can be reached from B via a metapath, there has to be at least one edge-dominant metapath, say M', from B to C. However, none of the edges in M' is a bridge, so for each edge e in M', there is at least one metapath $M''(B, C)$ in $S' = \langle X, E \backslash e \rangle$. However, since M' is edge-dominant, $M'' \not\subset M'$. Since every metapath can be reduced to an edge-dominant metapath, it follows that M'' can be reduced to an edge-dominant metapath that is distinct from M', which proves the result.   □

In the following, we use the notation $B \to C$ to denote that there is at least one metapath $M(B, C)$ from B to C.

THEOREM 3.3. *Given a bridge b between sets B and C in a metagraph:*

1. *For any subset D of X such that $B \to D$ and $D \to C$, b is a bridge between either B and D or between D and C.*
2. *If M_1 and M_2 are two metapaths from B to C, then $b \in M_1 \cap M_2$ and thus no two metapaths from B to C are edge-disjoint.*
3. *For any two sets B', C' such that $B' \to C'$, and $B' \subseteq B, C \subseteq C', b$ is a bridge between B' and C'.*

PROOF. (Part 1; proof by contradiction) Assume that b is not a bridge between either B and D or D and C. This implies that in the metagraph $S_1 = \langle X, E \backslash \{b\} \rangle$ we still have $B \to D$ and $D \to C$, and thus $B \to C$ (by the transitivity of '$\to$'). This in turn implies that b is not a bridge between B and C.

(Part 2) Since b is a bridge between B and C, it must occur in every metapath from B to C, and thus $b \in M_1 \cap M_2$.

(Part 3; proof by contradiction) Note that a metapath from B' to C' is also a metapath from B to C. Thus, if there is no bridge between B' and C', then there must be at least one metapath from B' to C' (and hence from B to C) in $S_1 = \langle X, E \backslash b \rangle$, which implies that b is not a bridge between B and C. □

Since the adjacency matrix of a metagraph and its closure can be used to find all the metapaths $M(B, C)$ between two sets of elements, these metapaths can be examined to identify bridges as well. Clearly, if any two of the metapaths are edge-disjoint, then by the above theorem, there is no bridge between the given set pair (B, C). If this is not the case, then, find the set of edges D such that every edge e in D occurs in every metapath $M(B, C)$ (by comparing the path(t) components of every pair of metapaths). Every such edge is then a bridge between B and C.

Chapter 4

METAGRAPH TRANSFORMATIONS

So far, we have considered a variety of features of a metagraph, where these features are specified in terms of the metagraph structure as given. However, there are many situations where it may be desirable to transform the given structure of a metagraph into another form that more effectively discloses certain structural features and/or facilitates certain analyses. In this chapter, we explore the transformation of a metagraph from one form to another that provides a different view of the system and/or relationships described by the metagraph.

There are several benefits of supporting views of systems. These include improved focus on relevant elements and relationships among them, logical independence between views and their basis, customization of views for different users of the system and modeling tool, and information sharing at different levels of abstraction.

In this chapter, we discuss three specific types of views that can be used to focus attention of different aspects of a large system. First, we describe the projection operation, which describes the relationships among a specific subset of the generating set of a base metagraph. Then, we discuss the pseudo-dual of metagraph, and finally, we discuss the element flow metagraph.

1. HIERARCHICAL ABSTRACTION USING PROJECTION

When a metagraph has a large number of elements and edges, the visualization benefits of the metagraph can be reduced by the resulting complexity, including the difficulty of rendering the visualization of so many relationships on a two-dimensional surface. In such situations, it is useful to be able to focus attention on a smaller set of elements and the relationships between them. The notion of a "view" has been used in the database literature to achieve such an effect in a complex, multiuser database (Date, 1995). Our approach to the definition of a view of a metagraph is through the use of a projection of the metagraph into a simpler metagraph.

DEFINITION 4.1. Given a metagraph $S = \langle X, E \rangle$ and $X' \subseteq X$, a metagraph $S' = \langle X', E' \rangle$ is a *projection* of S over X' if:

1. For any $e' = \langle V', W' \rangle \in E'$ and for any $x' \in W'$ there is a dominant metapath $M(V', \{x'\})$ in S, and
2. For every $x' \in X'$, if there is any dominant metapath $M(V, \{x'\})$ in S with $V \subseteq X'$, then there is an edge $\langle V', W' \rangle \in E'$ such that $V = V'$ and $x' \in W'$.
3. No two edges in E' have the same invertex.

The third condition simplifies the projection by minimizing the number of edges in it; it also allows the projection to be unique. Note that a projection does not show all relationships between the elements in the projection's generating set, only those that require minimal sets of inputs for each outvertex element. For instance, if there are two edges $\langle \{a\}, \{b\} \rangle$ and $\langle \{a, c\}, \{b\} \rangle$ in the given (or base) metagraph, then only the former edge appears in a projection that includes all these elements. In other words, the projection identifies only the necessary sets of elements for computing each element. While broader definitions have several merits and it may be useful to identify other views in some situations, it is difficult to operationalize such broader classes of views in general; that is, some of the structures that result from such broader definitions can be misleading.

We can illustrate the projection operation by an example. Consider the metagraph in Figure 4.1, and consider the projection over the subset $X' = \{x_1, x_2, x_6, x_7, x_8\}$ of its generating set. The projection, which appears in Figure 4.2, consists of three edges, each of which is a dominant metapath in S, and the vertices of which are contained in X'. No elements in $X \setminus X' = \{x_3, x_4, x_5\}$ appear in the projection.

The purpose of the projection is to provide a highlevel view of the metagraph that hides certain details. The projection hides certain details, in the

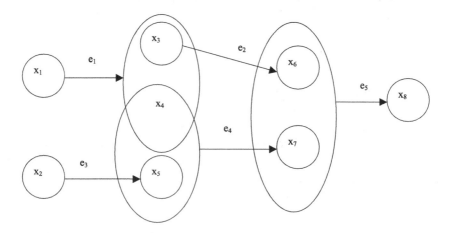

Figure 4.1. An example metagraph.

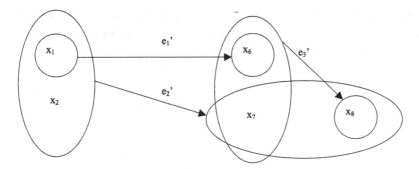

Figure 4.2. Projection of Figure 4.1 over $X' = \{x_1, x_2, x_6, x_7, x_8\}$.

sense that if an edge $e' = \langle V', W' \rangle$ appears in S', then it is possible to reach W' from V' in S, but there may be several other intermediate elements in $X \setminus X'$ that are also traversed. For example, in Figure 4.2 we can see that it is possible to calculate x_6 given only x_1, and the fact that x_3 is an intermediate variable is hidden from the person viewing the projection, since $x_3 \in X \setminus X'$. In addition, the fact that x_4 is also calculated in the process (by e_1) is hidden from the user, because $x_4 \in X \setminus X'$.

The advantage of a projection is that it may disclose relationships that are implicit in the original metagraph but are not easy to see because of the size and complexity of the original metagraph. The relationship, represented by e_1' between x_1 and x_6 is one example. Another example is provided by e_2', which represents the invocation of a metapath $M(\{x_1, x_2\}, \{x_7, x_8\}) = \{e_1, e_2, e_3, e_4, e_5\}$. It may not be clear from Figure 4.1 that this is a dominant metapath for the calculation of both LO and TL. The third edge in the projection, e_3', is easily discernible from the original metagraph, since it is simply e_5.

Because each edge in a projection corresponds to one or more metapaths in the original metagraph, it is useful to establish the set or sets of edges in the original metagraph that corresponds to each edge in the projection. This is called the composition of the projected edge and is defined as follows:

DEFINITION 4.2. Given a metagraph $S = \langle X, E \rangle$ and its projection $S' = \langle X', E' \rangle$ along $X' \subseteq X$, the *composition* $C(e')$ of an edge $e' \in E'$ is the set of metapaths in E that correspond to e'.

Thus, in Figures 4.1 and 4.2, $C(e_1') = \{\{e_1, e_2\}\}$, $C(e_2') = \{\{e_1, e_2, e_3, e_4, e_5\}\}$ and $C(e_3') = \{\{e_5\}\}$. We note that a composition is not a set of edges, but a set of sets of edges, because there may be more than one metapath in S corresponding to an edge in S'.

An interesting property of projection and composition as we have defined them is that they are unique and unambiguous, as stated in the following theorem (from Basu, Blanning and Shtub, 1997):

THEOREM 4.1. *The projection of a metagraph along a given subset of its generating set and the composition of each of its edges is unique.*

PROOF. Let $S = \langle X, E \rangle$ be the metagraph and let S', S'' be two projections of S over X', with $e' = \langle V', W' \rangle \in E'$ and $e' \notin E''$. Since $e' \in E'$, there exists a dominant metapath in S from V' to W'. Since $e' \notin E''$ and S'' is also a projection of S, there must be an edge $e'' = \langle V'', W'' \rangle \notin E''$ with either $V' \subset V''$ and $W' \subseteq W''$ or $V' \subseteq V''$ and $W' \subset W''$ (which follows from Definition 4.5). However, if $V' \subset V''$, then e'' violates condition 1 of Definition 4.6 for all $x \in W'$, and if $W' \subset W''$, then e' violates part 3 in Definition 4.6. Thus, $V' = V''$ and $W' = W''$, which implies that $e' = e''$, a contradiction that proves our result.

Now consider any $e' = \langle V', W' \rangle \in E'$. There is at least one metapath $M(V', W')$ in S from $V' \subseteq X'$ to $W' \subseteq X'$. Consider the set of all such metapaths $\{M, (V', W'), M_2(V', W'), \ldots\}$. Since there can be only one such set, $C(e')$ is unique. Therefore, for each e', $C(e')$ is unique.   □

An interesting property of projections that does not apply to metagraphs in general, is that in a projection, there can be no simple paths of any length between two elements unless there is an edge in the projection connecting the elements. The following theorem (also from Basu, Blanning and Shtub, 1997) states this formally:

THEOREM 4.2. *Given a metagraph $S = \langle X, E \rangle$ with adjacency matrix A' and closure A^*, and its projection $S = \langle X', E' \rangle$ over $X' \subseteq X$ with adjacency matrix A' and closure A'^*, $a'_{ij} = \phi$ iff $a'^*_{ij} = \phi$ for any $x_i, x_j \in X'$.*

PROOF. First, we show that if $a'^*_{ij} = \phi$, then $a'_{ij} = \phi$. This follows since $a'_{ij} \subseteq a'^*_{ij}$.

Next we prove by contradiction that if $a'_{ij} = \phi$, then $a'^*_{ij} = \phi$. Let i, j be such that $a'_{ij} = \phi$, but $a'^*_{ij} \neq \phi$. Then there must be at least one simple path, p, from x_i to x_j in S' consisting of a sequence of edges $\langle e'_{ij}, \ldots, e'_{ik} \rangle$ with each such $e'_{ij} \in E'$, $k = 1, \ldots, K$. The coinputs of x_i in this path are all elements of X'. Thus, there is a metapath from $x_i \cup \{x \mid x \text{ is a coinput of } x_i \text{ in } p)\}$ to x_i in S' (and also in S). From the definition of projection (Definition 4.6), if there is a metapath in S from some $V \subseteq X'$ to some $W \subseteq X'$, then $a'_{mn} \neq \phi$ for all $x_m \in V$ and $x_n \in W$. Since $x_i \cup \{x \mid x$ is a coinput of x_i in $p)\} \subseteq X'$ and $x_i \in X'$, it follows that $a'_{ij} \neq \phi$ which contradicts the assumption. $\qquad\square$

This theorem illustrates that in the case of the projection of a metagraph, any questions of reachability in the projection metagraph can be answered using the A' matrix itself (which is based on the A^* matrix for the base metagraph). The reason is that any connectivity in the base metagraph between elements of X' is captured in some edge in the projection. Thus, there is no need to consider paths of length greater than one in the projection, and the closure of the A' matrix does not have to be computed.

In summary, we have shown that using projection, a metagraph can be transformed into a higher-level view in which certain elements and relationships are retained and others are deliberately hidden from the user. Also, several views may be constructed from a single metagraph, by selecting different elements in X'.

We next show how to construct the projection of a metagraph over a given set of elements, using the A^* matrix. As indicated by part 2 of the definition of a projection, this requires the identification of all dominant metapaths between these elements. A brute force approach is to examine all combination of triples in A^*, but this could be extremely inefficient, because it involves examining 2^θ combinations of triples, where θ is the number of triples in A^*. A more efficient procedure can be built using the following observations about the A^* matrix:

- Any metapath between elements in X' must be composed of paths corresponding to triples in a^*_{ij} such that both $x_i, x_j \in X'$.
- Any triple in a^*_{ij} with $x_i, x_j \in X'$ such that all the coinputs are in X', corresponds to a valid metapath between elements in X'.

The value of these observations is that they allow us to reduce the number of triples in the A^* matrix that need to be considered. In the procedure **Projbuild** below, the projection is obtained by first building a set L, each element of which is a set of triples from A^* that comprises a candidate edge in the projection. Then, L is reduced by identifying dominant metapaths, leading to a set L_0 in which each element l corresponds to an edge in the projection. The composition of l is a collection of edge-sets, each of which corresponds to one or more sets of triples in L.

Procedure Projbuild (S, X')

1. Reduce A^* by eliminating all rows and columns corresponding to elements that are not in X'. Let the resulting matrix be A^*_0.

2. Using A_0^*, identify all edges $e \in E$ such that $V_e \subseteq X'$, and create a set L of all such edges (since e corresponds to a candidate edge in the projection).

3. For each combination of triples $\{t_l, \ldots, t_n\}$ from A_0^*, if $\bigcup_i \text{coin}(t_i) \backslash \bigcup_i \text{coout}(t_i) \subseteq X'$, then add the set of edges corresponding to this set of triples as an element of L.

4. Construct the set L_0 from L by taking each element l of L (each element of L is a set of edges in the base metagraph), and augmenting it with its net inputs, and its outputs (i.e., each element of L_0 is a triple ($\{$net inputs of $l\}$, $\{$outputs of $l\}$, l)).

5. Reduce L_0 to identify dominant metapaths, as follows:
   (a) Eliminate each triple $i \in L_0$ that is subsumed by at least one other triple (i.e., i is subsumed by j if the latter corresponds to a subset of edges in i and the outputs of j that are in X' include all such outputs of i);
   (b) For any $i \in L_0$ such that $\exists j \in L_0$, $j \neq i$ and both the outputs and inputs of j are subsets of those of i, eliminate all the elements in the output of i that are also outputs of j (since the edges in i are not a dominant metapath for those elements); if there are no remaining outputs in the triple, drop the triple.

6. Each triple in L_0 now corresponds to a dominant metapath between elements in X'. For each set of two or more triples with the same inputs and outputs, combine them into a single triple whose third component (a set, each element of which is the third component of the original triples) is the composition of the corresponding projection edge.

7. If there are still multiple triples having the same input, in order to satisfy part 3 of Definition 4.6, combine these into a single triple whose output is the union of all the outputs of the component triples; the composition of the corresponding projection edge e' is formed by taking one edge-set from the composition (edges) of each component triple, and computing the union of these edge-sets.

8. Return L_0.

It can be shown that this procedure **Projbuild** always terminates with a valid projection of the metagraph on the specified element set. The following theorem states this formally:

THEOREM 4.3. *Procedure **Projbuild** always terminates with a valid projection of the metagraph on the specified element set.*

PROOF. The procedure examines all relevant combinations of triples in A^* that could form metapaths between any pair of vertices in X'. Thus, this procedure finds all valid metapaths within X'. Furthermore, since a_0^* is finite, this procedure always terminates.

To show that the procedure correctly forms the projection, first assume that there is an edge e in the projection that is not identified by the procedure. Clearly, e corresponds to at least one dominant metapath between $V(e)$ and $W(e)$. However, since **Projbuild** finds all metapaths between elements in X', e must be included in L. Thus the only possibility is that e is dropped from L_0 at some stage. However, none of the reductions in Step 6 removes a triple that is essential for the projection, so e cannot be eliminated. Thus we have a contradiction.

Now assume that e is identified by the procedure, but is not in the projection. Clearly, because every element in L corresponds to a valid metapath within X', the only possibility is that e is not dominant. However, Step 6 eliminates all such metapaths, and Steps 7 and 8 merge all metapaths with common inputs. Thus e must be in the projection, which is a contradiction. ☐

Furthermore, the computational complexity of the procedure is quite manageable in most cases. As mentioned earlier, a naive approach would be to find all metapaths within the detailed metagraph, and then eliminate all those having inputs outside X'. In **Projbuild**, the first step eliminates a possibly large number of triples from contention. To appreciate this, note that even if the detailed metagraph is on a large generating set, most projections (if they are to serve a useful purpose for visualization and planning) are over a relatively small set; otherwise the projection would be very large and cluttered, and thus not much better than the detailed metagraph itself. Beyond this, the procedure involves examination of all combinations of the remaining triples, the complexity of which is exponential in the number of triples. Thus, the computational complexity of the procedure depends primarily upon the density of the relevant portion of the base metagraph (i.e., the number of triples in A_0^*), which may not be large even if the overall metagraph is very large.

So far, we have considered views of a single metagraphs. In general, however, there may be multiple related metagraphs in a given problem context. We have discussed the combination of metagraphs in an earlier chapter. We now examine whether views of multiple metagraphs can also be combined.

Consider a situation where there are two distinct metagraphs over possibly overlapping generating sets. If the metagraphs are $S_1 = \langle X_1, E_1 \rangle$ and $S_2 = \langle X_2, E_2 \rangle$, then the new metagraph, which we will call the sum of S_1 and S_2, will be $S_{12} = S_1 + S_2 = \langle X_1 \cup X_2, E_1 \cup E_2 \rangle$. If $X_1 \cap X_2 \neq \phi$, then S_{12} may contain simple paths and metapaths that are not entirely within either S_1 or S_2. To see this, consider two projections S_1' and S_2' of S_1 and S_2 respectively, where $S_1' = \langle X_1', E_1' \rangle$ is the projection of S_1 over $X_1' \subseteq X_1$ and $S_2' = \langle X_2', E_2' \rangle$ is the projection of S_2 over $X_1' \subseteq X_2$. If we combine the two views, we get a metagraph $S_1' + S_2' = \langle X_1' \cup X_2', E_1' \cup E_2' \rangle$. An interesting question is whether any information about relationships between elements of $X_1' \cup X_2'$ that existed in S_{12} (and thus in the projection of S_{12} over $X_1' \cup X_2'$) are lost in this process. That is, we would like to know whether $S_1' + S_2'$ contains the same information as the metagraph $S_{12}' = \langle X_1' \cup X_2', E_{12} \rangle$, which is the projection of S_{12} over $X_1' \cup X_2'$.

In order to do this we first elaborate on the notion of dominance discussed in a previous chapter, as follows:

DEFINITION 4.3. A metapath $M(B, C)$ in a metagraph *dominates* another metapath $M'(B', C')$, if $B \subseteq B'$ and $C' \subseteq C$.

DEFINITION 4.4. A metagraph S *dominates* another metagraph S' if every metapath in S' is dominated by some metapath in S.

Using these concepts, the relationship between S_{12}' and $S_1' + S_2'$ can be characterized in terms of the following theorem:

THEOREM 4.4. *Consider two metagraphs $S_1 = \langle X_1, E_1 \rangle$ and $S_2 = \langle X_2, E_2 \rangle$, along with their sum $S_{12} = \langle X_1 \cup X_2, E_1 \cup E_2 \rangle$. For some $X_1' \subseteq X_1$ and $X_2' \subseteq X_2$, let $S_1' = \langle X_1', E_1' \rangle$ and $S_2' = \langle X_2', E_2' \rangle$ be projections of S_1 and S_2, respectively. Also, let $S_{12}' = \langle X_1' \cup X_2', E_{12}' \rangle$ be a projection of S_{12} over $X_1' \cup X_2'$. Then S_{12}' dominates $S_1' + S_2'$.*

PROOF. We show that S_1' is dominated by S_{12}'. Consider an edge $e = \langle V, W \rangle \in E_1'$. Then there is a dominant metapath $M(V, W) = \{e_1, \ldots, e_n\} \subseteq E_1$ in S_1. Thus $M \subseteq E_1 \cup E_2$, and $V, W \subseteq X \subseteq (X_1 \cup X_2)$, and thus M is also a metapath in S_{12}. Thus, there is an edge $e^* = \langle V_1, W_1 \rangle$ in S_2' that dominates e (that is, $V_1 \subseteq V$ and $W \subseteq W_1$).

Similarly, S_2' is also dominated by S_{12}', and thus $S_1' + S_2'$ is dominated by S_{12}', and the result follows.   □

Note that the converse does not have to hold. That is, there may be some edges in S_{12} that are not dominated by any edges in $S_1' + S_2'$. To illustrate this, consider the metagraph in Figure 4.3 as S_2.

Now, considering Figure 4.1 as S_1 and Figure 4.2 as S_2, we can construct the following related metagraphs:

1. We combine the two metagraphs to get $S_{12} = S_1 + S_2$. Then we project S_{12} over the set $X_1' \cup X_2' = \{x_1, x_2, x_4, x_5, x_6, x_7, x_8, x_9, x_{13}\}$ to get S_{12}'. This is shown in Figure 4.4.

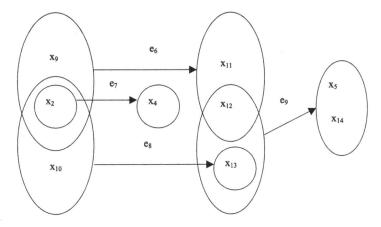

Figure 4.3. A second metagraph S_2.

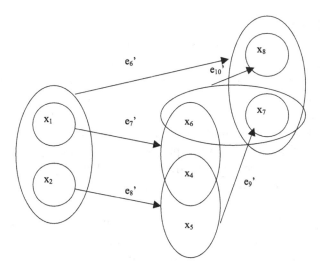

Figure 4.4. Projection S'_{12} of the joint metagraph.

2. We project S_1 over the set $X'_1 = \{x_1, x_2, x_6, x_7, x_8\}$, to get S'_1, and project S_2 over the set $X'_2 = \{x_9, x_2, x_4, x_5, x_{13}\}$ to get S'_2. Then we combine these two projections as $S'_1 + S'_2$. This is shown in Figure 4.5.

We can see from Figures 4.4 and 4.5 that S'_{12} dominates $S'_1 + S'_2$. For example, some edges, such as $\langle\{x_6, x_7\}, \{x_8\}\rangle$, appear in both S'_{12} and $S'_1 + S'_2$. It appears as e'_{10} in S'_{12} and e'_3 in $S'_1 + S'_2$. In this case the compositions of the relevant edges are the same: $C(e'_1) = C(e'_3) = \{\{e_5\}\}$. On the other hand, $e'_4 = \langle\{x_2, x_9, x_{13}\}, \{x_5\}\rangle$ in $S'_1 + S'_2$ does not appear in S'_{12}, but it is dominated by the edge $e'_8 = \langle\{x_2\}, \{x_5, x_4\}\rangle$ in S'_{12}. In addition, there are edges in S'_{12}

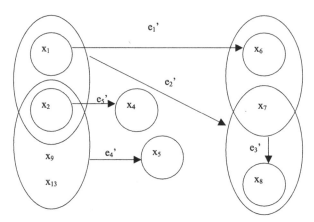

Figure 4.5. Combination of the projections $S_1' + S_2'$.

that do not dominate any edge in $S_1' + S_2'$. An example is $e_1' = \langle\{x_1\}, \{x_6, x_4\}\rangle$. Although the edge $e_1' = \langle\{x_1\}, \{x_6\}\rangle$ is found in $S_1' + S_2'$, there is no edge in $S_1' + S_2'$ with x_1 in its invertex and x_4 in its outvertex.

A possible negative consequence of integrating high-level views of a model base is that information may be lost. For instance, in the metagraph $S_1' + S_2'$, we find that the fact that x_4 is reachable from x_1 in the combined metagraph S_{12} is lost. This leads to the question of whether there are any conditions under which no information is lost in projecting two metagraphs and summing the projections (i.e., other than the trivial cases where $X_1' = X_1$, $X_2' = X_2$ or $X_1 \cap X_2 = \phi$). In other words, we have seen that every metapath in $S_1' + S_2'$ is dominated by a metapath in S_{12}'. The question then can be stated as whether there are any conditions under which the converse is true as well – that is, every metapath in S_{12}' is also dominated by a metapath in $S_1' + S_2'$. To do this we first establish a term for mutual dominance.

DEFINITION 4.5. Two metagraphs S_1 and S_2 are *equivalent* if they each dominate the other.

Note that equivalence is not the same as equality. That is – the edges in the equivalent metagraphs need not be the same. The difference will be illustrated below.

A sufficient condition for equivalence of S_{12}' and $S_1' + S_2'$ is that the intersection of the generating sets is contained within the intersection of the projection sets – that is, $X_1 \cap X_2 \subseteq X_1' \cap X_2'$. Because it is always true that $X_1' \cap X_2' \subseteq X_1 \cap X_2$ this condition is the same as $X_1' \cap X_2' = X_1 \cap X_2$. The following theorem states this formally:

THEOREM 4.5. *Consider two metagraphs* $S_1 = \langle X_1, E_1 \rangle$ *and* $S_2 = \langle X_2, E_2 \rangle$, *along with their sum* $S_{12} = \langle X_1' \cup X_2', E_1' \cup E_2 \rangle$. *For some* $X_1' \subseteq X_1$ *and* $X_2' \subseteq X_2$, *with* $X_1 \cap X_2 = X_1' \cap X_2'$, *let* $S_1' = \langle X_1', E_2' \rangle$ *and* $S_2' = \langle X_2', E_2' \rangle$ *be projections of* S_1, *and* S_2, *respectively. Also, let* $S_{12}' = \langle X_1' \cup X_2', E_{12}' \rangle$ *be a projection of* S_{12} *over* $X_1' \cup X_2'$. *Then* S_{12}' *is equivalent to* $S_1' + S_2'$.

PROOF. From Theorem 4.4, we know that S_{12} dominates $S_1' + S_2'$. We now show that under the stated condition, for any edge in S_{12}, there is a metapath in $S_1' + S_2'$ that dominates it.

Let $M(A, B)$ be a metapath from A to B in S_{12}, such that M is in the composition of some edge (e') in S_{12} from A to B. Partition M into two sets of edges M_1 and M_2 such that $M_1 \subseteq E_1$ and $M_2 \subseteq E_2$.

Consider M_1. Since M is a metapath between two sets of elements in $X_1' \cup X_2'$, every element x in netin(M_1) must be either (1) in X_1', or (2) in $X_1 \cap X_2$. The reason for this is that every net input of M is in $X_1' \cup X_2'$, and thus, all those elements in netin(M_1) that are not themselves in X_1' must be reachable from $X_1' \cup X_2'$ using edges in M_2. But the only elements in all such edges that are also in X_1 are those in $X_1 \cap X_2$. However, by the condition stated in the theorem, this implies that all the elements in case (2) above are in $X_1' \cap X_2'$. This, together with case (1), implies that all the elements in netin(M_1) are in X_1'. Similarly, all the elements in netin(M_2) are in X_2'.

Every element in B must be in the outvertex of some edge in M_1, M_2, or both. Partition B into B_1 and B_2 as follows:

$$ B_1 = B \cap \left(\bigcup_{e_i \in M_1} W_i \right), \qquad B_2 = B \cap \left(\bigcup_{e_j \in M_2} W_j \right). $$

By definition, M_1 is a metapath in S_1 from netin(M_1) to $\bigcup_{e_i \in M_1} W_i$. It follows that M_1 is a metapath from netin(M_1) to B, in S_1, and since both the source and target of the metapath are in X_1', there is an edge in S_1' between these two sets as well. Similarly for M_2 and B_2 in S_2. Thus, in $S_1' + S_2'$, there is a metapath from A to B composed of edges in $M_1 \cup M_2 = M$. Because this is true for every metapath M in S_{12}', the result follows. $\qquad \square$

The condition $X_1 \cap X_2 = X_1' \cap X_2'$, did not hold in the previous example, because $X_1 \cap X_2 = \{x_2, x_4, x_5\}$ but $X_1' \cap X_2' = \{x_2\}$. Thus there were elements common to the generating sets of metagraphs that were not common to the sets of elements over which the two metagraphs were projected.

By changing the elements over which the projections are constructed, we can illustrate what happens when the condition is satisfied. For instance, if we project S_1 over a new $X_1' = \{x_1, x_2, x_4, x_5, x_7\}$, we get a new $X_{12}' = X_1 \cap X_2 = X_1' \cap X_2' = \{x_1, x_4, x_5\}$ – that is, all of the elements that are found in the union of the generating sets of S_1 and S_2 are also in the sets of elements over which S_1 and S_2 are projected. The union of the sets over which the metagraphs are projected is $X_1' \cup X_2' = \{x_1, x_2, x_4, x_5, x_7, x_9, x_{13}\}$. In this case, as before, S_{12}' dominates $S_1' + S_2'$; however, the converse is also true, i.e., $S_1' + S_2'$ domina-tes S_{12}'.

It is also important to note that equivalence (i.e., mutual dominance) does not necessarily imply equality. The metagraphs in Figures 4.6 and 4.7 are clearly not the same, in part because of the existence of edge $e_4' = \langle\{x_2, x_9, x_{13}\}, \{x_5\}\rangle$ in Figure 4.7. However, this edge does not destroy the

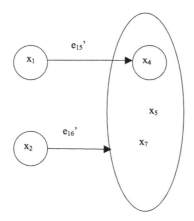

Figure 4.6. The second projection of S_{12} over $\{x_1, x_2, x_4, x_5, x_7\}$.

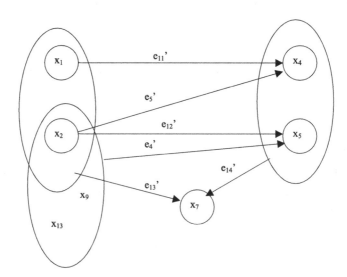

Figure 4.7. Sum of the projections in Figures 4.3 and 4.6.

equivalence of S'_{12} and $S'_1 + S'_2$, since it is dominated by e'_{16} in Figure 4.6. In addition, the edge $e'_1 = \langle\{x_1, x_2\}, \{x_7\}\rangle$ in Figure 4.7 does not appear in S'_{12}, but it is dominated by the edge e'_{16} in Figure 4.6.

Although S'_{12} (Figure 4.6) and $S'_1 + S'_2$ (Figure 4.7) are equivalent, there is a difference between them: S'_{12} seems simpler than $S'_1 + S'_2$. One reason for this is that a projection such as S'_{12} only contains dominant metapaths, whereas the sum of two projections such as $S'_1 + S'_2$, need not. For example, in Figure 7.17, the edge $e'_4 = \langle\{x_2, x_9, x_{13}\}, \{x_5\}\rangle$ is in $S'_1 + S'_2$, but not in S'_{12}, because it is dominated by e'_{12}. Another reason for the simplicity of S'_{12} is the requirement that no two edges can have the same invertex. Thus, edge e'_{16} in S'_{12} corresponds to three edges – e'_5, e'_{12}, and e'_{14} – in $S'_1 + S'_2$. As before, there is no such simplifying requirement for the sum of two projections, only for a single projection.

In summary, we have seen that there is a simple criterion for the integration of two views that avoids the misleading impression that a calculation cannot be performed (e.g., that expected service life cannot be calculated from the design variables) when in fact the calculation can be performed. The requirement is that all variables common to the two sets of calculations (i.e., the life cycle costing calculations and the cost estimating relationships) also be in both of the sets of variables used to construct the higher-level views. We have also seen that two views can be equivalent without being identical. In our second example, the sum of the higherlevel views of the life cycle costing calculations and of the cost estimating relationships was equivalent to a single view constructed directly from both sets of calculations.

At the same time, the ways in which these relationships were presented to the user were different in two respects. First, the sum of the high-level views contained redundant information, in the form of dominated metapaths. For example, the sum of the views disclosed that miles driven, fuel cost, and annual cost of preventive maintenance are sufficient to calculate annual operation and support cost, but it also disclosed that miles driven is sufficient to calculate annual operation and support cost. The second difference is that several relationships in the sum of the higher-level views may appear as only one relationship in the direct view of both sets of calculations. For example, the sum of views contained separate relationships between miles driven and service life, miles driven and annual operations and service cost, and miles driven and life cycle operation and support cost. In the direct view this is presented to the users as a single relationship: miles driven is sufficient to calculate all three variables. Thus, the users can easily see that several of the variables of interest to them are determined by a single variable, miles driven.

In cases where $X_1 = X_2 = X$ (i.e., the two base metagraphs are defined on the same generating set), Theorem 4.5 is not very useful. However, the following condition can still be used to check the base metagraphs before combining their projections:

COROLLARY. *Consider all elements $y \in [(X_1 \cap X_2)\backslash(X'_1 \cap X'_2)]$. Then, if there are no two distinct elements $z_1, z_2 \in (X_1 \cup X_2)$ such that there is a simple path from z, to y as well as a simple path from y to z_2 in $S_1 \cup S_2$, then S'_{12} is equivalent to $S'_1 + S'_2$.*

The proof of this corollary to is essentially the same as that for Theorem 4.5, because under the stated condition, all the elements in netin(M_1) are still either in X'_1 (or in $X'_1 \cap X'_2$, and thus in X'_1).

2.   THE INVERSE METAGRAPH

The inverse metagraph is a representation in which the generating set is made up of edges from the original metagraph, and where the edges correspond to combinations of elements from the original metagraph's generating set. Thus, the inverse metagraph is intuitively like the dual of a simple or directed graph. As with these simple graph constructs, one could construct a dual metagraph by simply transposing the incidence matrix of the original metagraph. However, the semantics of the resulting dual structure are different (in that the invertex of each edge represents a disjunction, rather than a conjunction, as in the primal metagraph). That is why we define the inverse metagraph, which provides the necessary "edge-centered" representation while retaining the conjunctive form of the edges.

DEFINITION 4.6. Given a metagraph $S = \langle X, E \rangle$, its *inverse* $T = \langle X', E' \rangle$ is a metagraph such that $X' = E \cup \{\alpha, \beta\}$, where α denotes the external source, β denotes the external target, and $e' \in E'$ iff in the primal metagraph, the primal elements corresponding to e' are in the outvertex of the primal edges in $V_{e'}$ and also in the invertex of the primal edges in $W_{e'}$. In addition (1) all pure inputs (i.e., elements that are not in any outvertex) in the primal appear in the inverse metagraph as edges from α, and (2) all pure primal outputs (i.e., elements that are not in any invertex) appear in the inverse metagraph as edges to β.

For example, the inverse of the metagraph in Figure 4.8 is the metagraph in Figure 4.9. Since both representations are metagraphs, properties such as paths, metapaths, cycles and bridges can be applied to the inverse metagraph just as to the primal, and the same algebraic operations and procedures can be applied to both. Thus, the inverse metagraph provides a complementary visual representation of a system that still supports metagraph analysis.

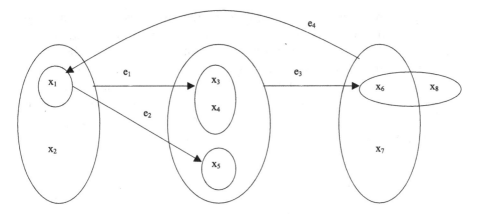

Figure 4.8. A primal metagraph.

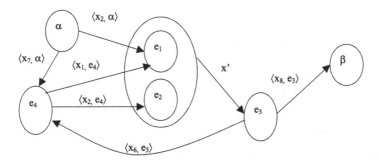

Figure 4.9. The inverse metagraph for Figure 4.8.

The inverse of a given metagraph can be generated from its incidence matrix G. This is a matrix whose rows correspond to the elements in the metagraph's generating set and whose columns correspond to the edges in the metagraph. There is a "+1" (resp., "−1") entry whenever the row element is an output (resp., input) of the edge corresponding to the column; all other entries are null.

The inverse metagraph can be constructed using the following procedure:

Procedure Inverse

1. For each column j of G, form all combinations of columns k such that $g_{ij} = -1$ and $g_{ik} = +1$, selecting no more than one column from each such row, and create an edge with each of the "+1" column indices in its invertex and each of the "−1" column indices in its outvertex. Label the edge with the set of row–column index pairs used to construct the edge

(e.g., if entries g_{ij} and g_{mn} are used for the invertices, and column p is used for the outvertex of edge y, then label $(y) = \{\langle x_i, e_j \rangle, \langle x_m, e_n \rangle\}$).

2. If a set of two or more edges have the same invertices and labels, then replace all these edges with a single edge having the same invertex and label, and the union of all the component outvertices in the outvertex.

3. For each row j in G that has only -1 (resp., $+1$) entries, create a single edge to (resp., from) the column entries from (resp., to) α (resp., β), with the label $\{\langle x_i, \alpha \rangle, \langle x_j, e_k \rangle\}$, for each column e_k with a "$+1$".

For the metagraph in Figure 4.1, the matrix G is as follows:

G	e_1	e_2	e_3	e_4
x_1	-1	-1	0	$+1$
x_2	-1	0	0	0
x_3	$+1$	0	-1	0
x_4	$+1$	0	-1	0
x_5	0	$+1$	-1	0
x_6	0	0	$+1$	-1
x_7	0	0	0	-1
x_8	0	0	$+1$	0

and the algorithm would proceed as follows:

1. The four edges $\langle \{e_4\}, \{e_1\} \rangle$ (with label $\langle x_1, e_4 \rangle$), $\langle \{e_4\}, \{e_2\} \rangle$ (with label $\langle x_2, e_4 \rangle$), $\langle \{e_1, e_2\}, \{e_3\} \rangle$ (with label $\langle x_3, e_1 \rangle, \langle x_4, e_1 \rangle, \langle x_5, e_2 \rangle$) and $\langle \{e_3\}, \{e_4\} \rangle$ (with label $\langle x_6, e_3 \rangle$), are identified.

2. The edges $\langle \alpha, \{e_1\} \rangle$ (with label $\langle x_2, \alpha \rangle$), $\langle \alpha, \{e_4\} \rangle$ (with label $\langle x_7, \alpha \rangle$), and $\langle \{e_1\}, \beta \rangle$ (with label $\langle x_8, e_3 \rangle$) are added.

The result is the inverse metagraph in Figure 4.2. Note that the inverse is equivalent to the original metagraph, in that the latter can be reconstructed given the former.

3.   THE ELEMENT FLOW METAGRAPH

Another operation, one that focuses attention on the flow of particular primal elements, is the transformation of a primal metagraph into its corresponding element flow metagraph (EFM). We first define the EFM, provide some intuition on its structure, specify an algorithm for generating it for a given metagraph, and then illustrate the algorithm with an example.

DEFINITION 4.7. Given a metagraph $S = \langle X, E \rangle$ and a specific subset X' of X, the *element flow metagraph* corresponding to X' is a metagraph $\overset{\bullet}{S} =$

$\langle X', F \rangle$ in which for each edge $f = \langle V_f, W_f \rangle \in F$, there exist edges $e_1, e_2 \in E$ such that

1. $V_f \subseteq V_{e_1}$,
2. $W_{e_1} \cap V_{e_2} = Z \neq \varnothing$, and
3. $W_f \subseteq V_{e_2}$.

The set of elements Z is the *flow content* on f through the (primal) edge pair e_1, e_2 from V_f to W_f. Since there could be several edge pairs $e_1, e_2 \in E$ corresponding to the same $f \in F$, we also define the *flow composition* $C(f)$ as the set of all such edge pairs representing flow on f.

Informally, this transformation is a simplified form of the primal in which the generating set consists of X', and each edge e' identifies what elements in $X \backslash X'$ flow from vertices in S containing the elements in $V_{e'}$ to elements in vertices containing the elements in $W_{e'}$. Thus, the edges represent a dependency between the elements in X'. However, this dependency is not of the type represented by the projection operator, which identifies metapaths between vertices. An edge e' in the EFM represents the fact that there is some edge in S whose invertex contains the elements in $W_{e'}$, and which requires input from some edge whose invertex contains the elements in $V_{e'}$.

For example, the EFM defined for the elements x_2, x_4 and x_7 from the metagraph in Figure 4.8 is shown in Figure 4.10. The composition of each edge is shown in the legend, where the content of each edge pair is included in parentheses. Once again, as with the inverse, the EFM can be analyzed using metagraph analytical procedures. For example, cycles can be identified and analyzed, metapaths can be identified to identify the scope of the impact of one set of resources upon another, and bridges can be used to identify critical element flows. However, since the edges in the EFM only represent direct element flows (across a single edge in the original metagraph), transitive element flows are captured only when the set X' covers all the edges in the original metagraph (in the sense that for $e \in E$, $X' \cap V_e \neq \varnothing$).

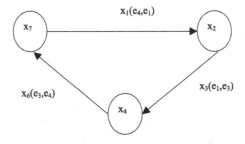

Figure 4.10. The element flow metagraph for Figure 4.9 with $X' = \{x_2, x_4, x_7\}$.

While the EFM could be used for any subset of X, it is particularly useful when X' is a separate type of element. For instance, in Chapter 7 we will show how this transformation is useful when X' represents a set of resources. It directly provides a view that represents the interaction between resources.

The EFM can be generated from a given metagraph by the following **Procedure EFM**, which uses the element task incidence matrix G. However, before describing this procedure, we need to define a new operator.

DEFINITION 4.8. Let A be an $m \times n$ matrix with row indices $x_1, \ldots, x_m$ and column indices $y_1, \ldots, y_n$. Let B be an $n \times p$ matrix with row elements $y_1, \ldots, y_n$ and column indices $z_1, \ldots, z_p$. Then the operation $A \otimes B$ results in an $m \times p$ matrix C with row indices $x_1, \ldots, x_m$ and column indices $z_1, \ldots, z_p$, such that

$$c_{ij} = \bigcup_{k=1,\ldots,n} a_{ik} \oplus b_{kj}$$

and

$$a_{ik} \oplus b_{kj} = \begin{cases} y_k & \text{if } a_{ik} = 1 \text{ and } b_{kj} = -1, \\ -y_k & \text{if } a_{ik} = -1 \text{ and } b_{kj} = -1, \\ 0 & \text{otherwise.} \end{cases}$$

Now, let G_1 be the sub-matrix of G corresponding to the elements in X' and G_2 be the sub-matrix of G corresponding to elements in $X \setminus X'$ that are not either pure inputs or pure outputs. Then the procedure is defined as follows:

Procedure EFM

1. Perform the operation $G_2 \otimes G_1^{\mathrm{T}}$ (G_1^{T} is the transpose of G_1); the result is a matrix R whose rows correspond to the non-terminal elements in $X \setminus X'$ and whose columns correspond to the elements in X'.
2. For each row r_i of R corresponding to element x_i, construct edges as follows:
   (a) for each edge t_a (resp., t_b) appearing as a positive (resp., negative) entry in any cells in r_i, create an invertex (resp., outvertex) consisting of all column indexes corresponding to the columns z_j such that $t_a \in r_{ij}$ (resp., $-t_b \in r_{ij}$);
   (b) combine each such invertex and outvertex pair as an edge. The flow composition of the edge is defined as $\{r_i \langle t_a, t_b \rangle\}$;
   (c) combine all edges with the same vertices into a single edge whose flow composition is the union of the flow composition of the component edges.

The procedure can be applied to the metagraph in Figure 4.8 to generate the EFM in Figure 4.10 for $X' = \{x_2, x_4, x_7\}$ as follows:

Step 1: Based on the partitioning of the incidence matrix with x_2, x_4, x_7 for G_1 and x_1, x_3, x_5, x_6 for G_1, the R matrix is as follows:

R	x_2	x_4	x_7
x_1	$-e_1$	0	$+e_4$
x_3	$+e_1$	$-e_3$	0
x_5	0	$-e_3$	0
x_6	0	$+e_3$	$-e_4$

Step 2: From the row for x_1, we get the edge $e' = \langle x_7, x_2 \rangle$ with flow composition $\{x_1 \langle e_4, e_1 \rangle\}$; similarly, we get the edges e'' and e''' from rows x_3 and x_6, respectively.

Note that there is no edge corresponding to the element x_5. This is because the edge that produces that element is not covered by X'. This illustrates why the covering condition stated earlier is important in order for an EFM to capture all element flows, both direct and indirect, among the elements of X'.

Chapter 5

ATTRIBUTED METAGRAPHS

As described thus far, metagraph edges are set-to-set mappings with no further information attached. However, it is possible to attach attributes to metagraph edges. In this chapter, we examine how both qualitative and quantitative attributes can be added to metagraph edges.

1.   QUALITATIVE ATTRIBUTES

A qualitative attribute is essentially a label that is added to each metagraph edge, in addition to the edge identifier (name) label. An example of such an attribute is a color, like 'Blue' or 'Red'. Since such attributes are only labels, they cannot be used in arithmetic operations. However, they can still be very useful in algebraic analysis of metagraph structure. At the simplest level, they can be used to partition the metagraph (e.g., into edges of the same color or type). They can also be used to constrain path or metapath selection.

The idea of adding qualitative attributes to edges is also well-established in traditional graphs and digraphs. However, the richer structure of a metagraph provides an interesting additional dimension, namely the separation of the visual and algebraic representation of such attributes. To see this, consider the examples in Figures 5.1. Both Figures 5.1(a) and 5.1(b) represent the same metagraph, with the edges assigned a color attribute. The first representation, in Figure 5.1(a), is visually more informative, and also familiar to users of attributed graphs and digraphs. On the other hand, the representation in Figure 5.1(b) shows each color attribute as an additional element of the generating set, and then included in the invertex of the edges to which it applies.

There are two important implications of this feature in metagraphs. First, while the edge representation is easier to interpret visually, the advantage of the vertex representation is that each attribute value needs to appear only once, and all edges with that attribute value are immediately identifiable (since that value is in their invertex). So in a large metagraph, finding all the blue edges is easy in the vertex representation. Of course, it may also be very difficult to draw this metagraph on a planar surface, once it has a large number of edges.

The second implication is that due to the semantic equivalence of the two representations, it is possible to use the edge representation in graphical visualization of attributed metagraphs, while using the vertex representation for

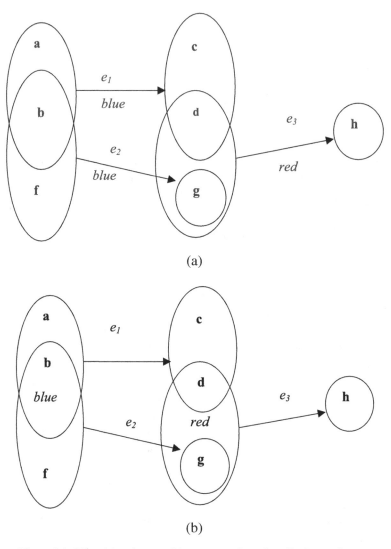

Figure 5.1. Edge (a) and vertex (b) representation of qualitative attributes.

analysis. The significance of this is that in the vertex representation, the attributes are treated simply as a subtype of element from the generating set. Thus, attributes are included in the adjacency matrix, and can be subject to the same types of analysis as any other element. In other words, the addition of qualitative attributes does not force any significant change to the basic algebraic constructs and operations in metagraphs, unlike other graph representations.

2.   QUANTITATIVE ATTRIBUTES

It is also possible to attach quantitative (numerical) attributes to metagraph edges. The purpose of this would be to allow certain calculations to be performed. For example, if the attributes are the costs of the tasks represented by the edges, then these attributes can be used to determine the total cost of the tasks appearing in a workflow. If the attributes represent the durations of the tasks, then they can be used to calculate the duration of the workflow in a fashion similar to the PERT/CPM calculations used in project management. If the attributes represent measures of performance, such as degrees of reliability or probabilities of non-failure, then they can be used to determine the performance of the workflow.

These attributes might also be combined. For example, if certain numerical attributes represent both time (i.e., activity durations) and cost, then these attributes might be combined to perform time/cost tradeoffs. If they represent either cost or duration along with probability of non-failure, then they might be used to determine the probability distributions of workflow cost or duration, depending on what will be done if a task represented by an edge should fail. In this paper we will focus on deterministic activity durations, and we will not consider time/cost tradeoffs.

3.   CONDITIONAL METAGRAPHS

One particular type of qualitative attribute in metagraphs that is used in a number of applications is a general proposition or proposition. A *proposition* is a statement that may be either true or false. If a proposition appears in the invertex of an edge, it must be true for the edge to be used in a metapath. Each edge may contain zero, one, or more assumptions, and each assumption may appear in one or more edges. Propositions appear in the generating set along with elements representing other types of variables. We distinguish metagraphs that contain propositions by the term conditional metagraph. We define a conditional metagraph as follows.

DEFINITION 5.1. A *conditional metagraph* is a metagraph $S = \langle X_p \cup X_v, E \rangle$, in which X_p is a set of *propositions* and X_v is a set of *variables*, and:

1. $\forall e' \in E, V_{e'} \cup W_{e'} \neq \varnothing$;
2. $X = X_v \cup X_p$ with $X_v \cap X_p = \varnothing$ such that $\forall p \in X_p, \forall e' \in E$, if $p \in W_{e'}$, then $W_{e'} = \{p\}$.

Note that a metagraph as defined in Chapter 2 is a specialization of a conditional metagraph in which $X_p = \varnothing$. Where significant, we can refer to such a metagraph as a simple metagraph.

Thus, conditional metagraphs must meet two constraints in addition to the requirement that the generating set be partitioned into variables and propositions. First, for each edge, at least one of the vertices must be nonempty, and the invertex and outvertex of each edge must be disjoint. Second, if an outvertex contains a proposition, then it cannot contain any other element.

The values of different propositions can be used to specify alternative contexts for a conditional metagraph. Thus, if a set of propositions P is true, another set Q is false, and the remaining propositions (i.e., $X_p \setminus (P \cup Q)$) are undetermined, then this knowledge can be used to simplify a conditional metagraph, so that only those edges that are valid in that context are retained. Specifically, the context of a conditional metagraph S with respect to P and Q, denoted $K(P, Q, S)$, is also a conditional metagraph in which (1) any proposition in P is deleted and (2) any edge containing a proposition in Q is deleted. If as a result any edge now has a null invertex or a null outvertex, then that edge is deleted as well. The resulting context metagraph will be a conditional metagraph because the undetermined propositions will remain as propositions in the context. The context operation is a useful abstraction on metagraphs, since it avoids the need to consider edges that cannot be used under the stated (Q) conditions.

The following definition of a context is a constructive definition specifying the simplification process.

DEFINITION 5.2. Given a conditional metagraph $S = \langle X_v \cup X_p, E \rangle$, a set of propositions $P \subseteq X_p$ that are known to be true and a set of propositions $Q \subseteq X_p$ that are known to be false, we define a *context* $K(P, Q, S)$ as a conditional metagraph derived from S as follows:

1. For any edge $e' \in E$ containing a proposition $p \in P$ simplify the edge by deleting p; if the resulting edge has a null in- or out-vertex, delete the edge;
2. For any edge $e' \in E$ containing a proposition $q \in Q$ in either vertex, delete the edge (only the edge and q are deleted, not the other elements in the edge's vertices).

In transforming a conditional metagraph into a context, the propositions whose truth values are known (i.e., $P \cup Q$) no longer appear and need not be considered in the model selection process. Thus, a context represents a simplified view of a model base that allows a user to consider only those models known to be relevant, and those variables and propositions whose values can be manipulated (e.g., in a sensitivity analysis). The larger the sets P and Q (i.e., the more specific the context), the simpler is the resulting conditional metagraph.

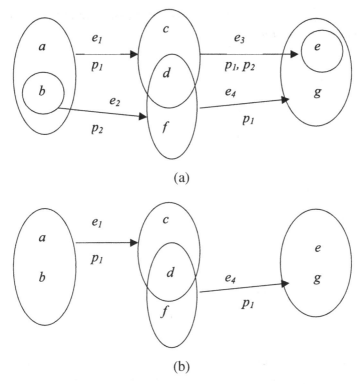

Figure 5.2. (a) A conditional metagraph; (b) the corresponding context metagraph.

An example of a context metagraph is shown in Figure 5.2. Starting with the conditional metagraph S in Figure 5.2(a), then $K(\{p_1\}, \{p_2\}, S)$ is the conditional metagraph in Figure 5.2(b).

Given a conditional metagraph, we can ask the following questions:

1. What propositions are associated with a given metapath?
2. Which of these propositions may be assigned values initially, and which will depend on the execution of edges in the metapath?
3. If we know at the start that certain propositions are true and that certain other propositions are false, how will that affect our decision analysis strategy?

In order to answer these questions, we define a conditional metapath as follows.

DEFINITION 5.3. Given a conditional metagraph $S = \langle X_v \cup X_p, E \rangle$ a source $B \subseteq X_v$ and a target $C \subseteq X_v$, a *conditional metapath* is a set of edges

$CM(B, C) = \{\ell'_\iota, \ \iota = 1, \ldots, L\}$, forming a metapath from $B \cup X_p$ to C. The *set of relevant propositions* is

$$\alpha = \left(\bigcup_{\iota=1}^{L} V'_\iota \right) \cap X_p.$$

This may be partitioned into two subsets, the *set of input propositions*

$$\beta = \left(\bigcup_{\iota=1}^{L} V'_\iota \setminus \bigcup_{\iota=1}^{L} W'_\iota \right) \cap X_p,$$

and the *set of intermediate propositions* is

$$\gamma = \left(\bigcup_{\iota=1}^{L} V'_\iota \right) \cap \left(\bigcup_{\iota=1}^{L} W'_\iota \right) \cap X_p.$$

Thus a conditional metapath establishes a relationship between two sets of variables, using whatever propositions are necessary to execute the necessary edges. The relevant propositions are those that appear in the metapath. The propositions appear in the invertex of at least one of the edges in the conditional metapath and it may appear in one of the outvertices as well. If a proposition does not appear in any of the outvertices, it is a member of the set of input propositions; otherwise, it is a member of the set of intermediate propositions.

The set of input propositions can be evaluated before any of the edges in the conditional metapath are executed. The intermediate propositions are evaluated based on some of the outvertex elements of edges in the metapath, once the values of those elements are known. The truth or falsity of any, other propositions – that is, any propositions not in the relevant set – will have no impact on the effectiveness of the conditional metapath in linking the source elements to the target elements.

3.1.   Projections in Conditional Metagraphs

In the previous chapter, we had introduced the notion of a projection operator, and showed how it could be used to construct hierarchical views of simple metagraphs. We now invoke the projection operator for conditional metagraphs, and then investigate the relationship between projections and contexts in the following subsection.

DEFINITION 5.4. Given a conditional metagraph $S = \langle X_v \cup X_p, E \rangle$, a *projection* of S over the set $X' \subseteq X_v$ is a conditional metagraph $N(X', S) = \langle X' \cup X_p, E' \rangle$ such that $e' \in E$ *iff* there is a dominant metapath from $V_{e'}$ to $W_{e'}$ in S.

We note two interesting features of this definition. First, the variables in the projection are limited to those in X', so none of the variables in $X \backslash X'$ appear in the projection. Some or all of the assumptions in X_p may appear in the projection; the user only specifies X', and the necessary assumptions are determined by the definition. Second; the projection (and therefore its adjacency matrix) represents all relationships among the variables in X' and $\alpha_{ij}'^* = \varnothing$ iff $\alpha_{ij}' = \varnothing$ (where A' is the adjacency matrix of the projection).

The first observation implies that the decision maker needs only to specify the relevant variables over which the projection is desired, and the operation then generates the relevant assumptions for each projected relationship. The second observation implies that use of projections does not require the computation of the closure of A' (i.e., A'^*), which saves some computational effort. The projection itself can be computed using the A^* matrix of the underlying metagraph. Although the complexity of the procedure is exponential in the number of triples in the relevant portion of the A^* matrix, the size of this portion depends upon the projection set (the elements in the generating set that define the projection). Since in practice this set is not large (otherwise, the benefit of constructing the projection is lost), the procedure is still practical.

Consider again the conditional metagraph illustrated in Figure 5.3. If we project this metagraph over the set of variables $X' = \{ADV, ECON, NI, UCOST, VOL\} \subset X_v$, the result is the conditional metagraph illustrated in Figure 5.4. Two propositions, *cadv* and *mkt*, do not appear in the projection. The reason is that *cadv* appears only in the invertex of e_2 and *mkt* appears only in the invertex of e_6, and neither of these edges is a member of a dominant metapath corresponding to the edge e_1'' in the projection.

We now consider the issue of combining projections and contexts. More specifically, we address the following questions:

1. Can a context be defined on a projection (and conversely, can a projection be defined on a context)?
2. If both transformations are to be applied on a base conditional metagraph, does the order matter (i.e., are these operations commutative)?

The answer to the first question is yes, for the simple reason that the result of both operations is a conditional metagraph, which is amenable to all the standard operations on conditional metagraphs. The answer to the second question is also yes, as evidenced by the following theorem:

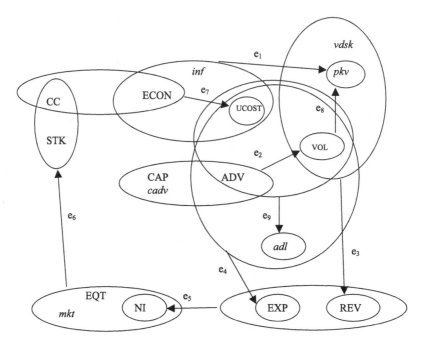

Figure 5.3. Conditional metagraph.

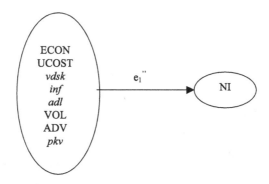

Figure 5.4. Projection of Figure 5.3 over $\{ADV, ECON, NI, UCOST, VOL\}$.

THEOREM 5.1. $N(X', K(P, Q, S)) = K(P, Q, N(X', S))$.

PROOF. Let $S_1 = K(P, Q, N(X', S))$, $S_2 = N(X', K(P, Q, S))$. The proof is by contradiction. That is, we show that it is not possible for an edge to be in S_1 and not in S_2, and vice versa.

1. Let $e \in S_1, e \notin S_2$. Since $e \in S_1$ $\exists$ an edge e' in $N(X', S)$ such that $in(e) \subseteq in(e')$, $out(e') \subseteq out(e)$ and $in(e')\backslash in(e) \subseteq P$;

$\Rightarrow \exists$ a metapath $M(in(e'), out(e))$ in S;

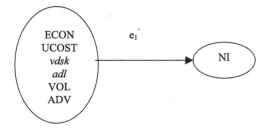

Figure 5.5. $N(\{ADV, ECON, NI, UCOST, VOL\}K(\{inf, pkv\}, \{cadv, mkt\}S)) = K(\{inf, pkv\},$ $\{cadv, mkt\}N(\{ADV, ECON, NI, UCOST, VOL\}, S)).$

$\Rightarrow \exists$ a metapath $M'(in(e), out(e))$ in $K(P, Q, S)$.

But since $e \notin S_2 \nexists$ a metapath from $in(e)$ to $out(e)$ in $K(P, Q, S)$, which is a contradiction.

2. Now let $e \in S_2, e \notin S_1$. Since $e \in S_1$, there is no edge e' in $N(X', S)$ such that $in(e) \subseteq in(e'), out(e') \subseteq out(e)$, and $in(e')\backslash in(e) \subseteq P$;

$\Rightarrow \nexists$ any metapath from $in(e')$ to $out(e)$ in S;
$\Rightarrow \nexists$ any metapath from $in(e)$ to $out(e)$ in $K(P, Q, S)$.

But, since $e \notin S_2$, it follows that $\exists$ a metapath from $in(e)$ to $out(e)$ in $K(P, Q, S)$, which is a contradiction, and thus the result follows. $\square$

We illustrate this commutativity property using our earlier example, by observing that if we construct the context metagraph for Figure 5.3 with $P = \{inf, pkv\}$, $Q = \{cadv, mkt\}$, and then project this conditional metagraph over $X' = \{ADV, ECON, NI, UCOST, VOL\}$, we get the conditional metagraph in Figure 5.5, which is also the result of defining the context $P = \{inf, pkv\}$, $Q = \{cadv, mkt\}$ on the projection of Figure 5.3 over X'.

3.2.   Connectivity and Redundancy

We now introduce two important properties of metagraphs. The first, connectivity and especially full connectivity, determines the ability of a metagraph to connect certain input variables to certain output variables. The second property is redundancy – that is, a determination of whether there is more than one way to connect an input to an output. We begin with some definitions.

DEFINITION 5.5. Given a conditional metagraph $S = \langle X_p \cup X_v, E \rangle$, any two sets $B \subseteq X_v$, and $C \subseteq X_v$, and R, a defined set of logic expressions over X_p, let $M(B, C, S)$ be the set of all edge-dominant metapaths from B to C. An *interpretation* $I(X_p, R)$ is an assignment of truth values to the propositions in X_p such that all the expressions in R evaluate to true. $P \subseteq X_p$ denotes the set of *true* propositions in $I(X_p, R)$ and $Q \subseteq X_p$ denotes the set of *false* propositions in $I(X_p, R)$.

To illustrate this, consider again the example metagraph in Figure 5.1. Let $B = \{a, c\}, C = \{d\}$, and R be the single logical expression $(p_1 \vee p_2)$. Then $M(B, C, S) = \{\langle e_1, e_2, e_3 \rangle, \langle e_1, e_4 \rangle\}$.

We note that a context metagraph $K(P, Q, S)$ corresponding to an interpretation $I(X_p, R)$ is a simple metagraph (i.e., it has no propositions), since $P \cup Q = X_p$.

DEFINITION 5.6. Given a conditional metagraph $S = \langle X_p \cup X_v, E \rangle$, any two sets $B \subseteq X_v$, and $C \subseteq X_v$, and R, a set of logic expressions defined over X_p:

1. B is said to be *connected* to C with respect to R if for some interpretation $I(X_p, R), |M(B, C, K(P, Q, S))| \geq 1$;
2. B is said to be *fully connected* to C with respect to R if for every interpretation $I(X_p, R), |M(B, C, K(P, Q, S))| \geq 1$.
3. B is said to be *non-redundantly connected* to C with respect to R if for every interpretation $I(X_p, R), |M(B, C, K(P, Q, S))| \leq 1$.

It follows from (2) and (3) above that B is *fully and non-redundantly connected* to C if $|M(B, C, K(P, Q, S))| = 1$ for every interpretation $I(X_p, R)$.

DEFINITION 5.7. A conditional metagraph $S = \langle X_p \cup X_v, E \rangle$, is *non-redundant* with respect to R, a set of logic expressions over X_p, if in the context metagraph $K(P, Q, S) = \langle X_p \cup X_v, E_k \rangle$ corresponding to any interpretation $I(X_p, R) \forall x \in X_v$ we have $|\{e \in E_k \mid x \in W_e\}| \leq 1$.

Informally, a metagraph is non-redundant if for every interpretation, each element is in the outvertex of at most one edge. In other words, there is at most one way to determine the value of the element in each interpretation, so there is no ambiguity.

Given the algebraic representation of a conditional metagraph in terms of its adjacency matrix A and closure, A^*, the set of metapaths from one set of elements B to another set of elements C can be identified using A^*. The procedure is as follows:

1. A^* is reduced to the rows corresponding to B and the columns corresponding to C (since all edge-dominant metapaths from B to C can be constructed from this sub-matrix);
2. For each interpretation, the corresponding reduced context metagraph is generated;
3. All valid metapaths from B to C in the context are then constructed for that interpretation using the procedure specified in Chapter 3.

The computational complexity of the above procedure depends upon the size of the proposition set X_p. In general, the number of possible interpretations of a given set of propositions and set R of expressions is exponential in N, the number of propositions (the worst case complexity is 2^N). However, in many practical situations, the number of possible interpretations will be sufficiently small to render the procedure feasible. For instance, in the context of process modeling, the number of interpretations of R corresponds to the number of alternate workflows for the represented process, which is not likely to be very large for typical business processes. Also, R may have some special structure that can be exploited to restrict the search for interpretations. For example, in process modeling, R may contain a number of expressions of mutual exclusivity between complementary literals (e.g., $X_p = \{$salaried, hourly, high-risk, low-risk$\}$ and $R = \{$(salaried $\oplus$ hourly), (high-risk $\oplus$ low-risk)$\}$, where $\oplus$ means exclusive disjunction).

Given a metagraph S with pure inputs PI and pure outputs PO, we can test whether PI is fully connected to PO in S as follows:

1. Let S be a simple metagraph. Then S is fully connected if there exists any $M(PI, PO)$;
2. Let S be a conditional metagraph, with a set of propositions X_p used as assumptions. Assume that all X_p are pure inputs. Then S is fully connected *iff* there exists a $M(PI, PO)$ for every possible interpretation of X_p. In effect, this implies that S is fully connected *iff* there exists a $M(PI, PO)$ even when all X_p elements are false.

In addition to case 2, let R be a set of Horn clause assertions on X_p. Again, S is fully connected *iff* there exists a $M(PI, PO)$ even when all X_p elements are false. The only additional feature here is that even when the truth value of all propositions in X_p are not explicitly known, the unknown propositions that are heads of clauses in R can be inferred.

Consider the union S_3 of two fully connected metagraphs S_1 and S_2 (i.e., $S_3 = S_1 \cup S_2$). Since PO_1 can be reached from PI_1 in all interpretations, and PO_2 can be reached from PI_2, then $PO_1 \cup PO_2$ is reachable from $PI_1 \cup PI_2$ in all interpretations. Then it follows that the pure inputs $PI_3 \subseteq PI_1 \cup PI_2$, and the pure outputs $PO_3 \subseteq PO_1 \cup PO_2$.

However it is important to realize that the pure inputs and pure outputs of the combined metagraph need not be $PI_1 \cup PI_2$ and $PO_1 \cup PO_2$ respectively. Therefore, it does not necessarily follow that S_3 is fully connected, as is demonstrated by the following example. Let S_1 consist of the edge $\{(a, b), (c, d)\}$ and S_2 consist of the edge $\{(d, f), (b, g)\}$. If $S_3 = S_1 \cup S_2$, then $PI_3 = \{a, f\}$ and $PO_3 = \{c, g\}$, and there is no metapath from PI_3 to PO_3. Thus, S_3 is not fully connected. Although neither S_1 nor S_2 is cyclic, S_3 is cyclic.

THEOREM 5.2. *Given two fully connected metagraphs S_1 and S_2, their union $S_3 = S_1 \cup S_2$ is also fully connected if it is acyclic.*

PROOF. Since S_3 is acyclic, its elements can be organized in a partial order based on the existence of simple paths between elements (i.e., p precedes q if there is a simple path from p to q). It follows that the elements in PI_3 are roots and elements of PO_3 are leaves of the precedence graph.

Consider an element x in PO_3 that is not reachable from PI_3 in some interpretation. Without loss of generality, assume that x is in PO_1. Since x is reachable from some subset of PI_1, say PI_{1x}, then it must be true that $PI_{1x} \backslash PI_3 \neq \varnothing$. Let $PI_{1x} \backslash PI_3 = Y$. Then Y consists of elements that were in PI_1, but are not in PI_3. Thus, each element of Y must be either an internal element of S_2 or a pure output of S_2. The precedence graph to each such element has to ultimately end with roots that are in PI_2, say PI_{2x}. If all these elements are in PI_3, then x is reachable from PI_3 and we are done. However, if $PI_{2x} \backslash PI_3 = Z \neq \varnothing$, then as before, these must be internal or pure output elements in S_1. Since S_3 is acyclic, Z is reachable from a subset of PI_1 say PI_{1x2} such that $PI_{1x2} \cap PI_{1x} = \varnothing$. Since both PI_1 and PI_2 are finite, these iterations must terminate, which proves the result. $\qquad\qquad\square$

This theorem provides a two-step test for whether the combination of two fully connected metagraphs is also fully connected, as follows:

1. Compute A^* and examine its diagonal elements. If all of the diagonal cells are empty, then full connectivity still holds because the metagraph is acyclic;
2. Test whether there is any cycle containing elements either from PI_1 and PO_2 or PI_2 and PO_1. If not, then full connectivity still holds.

If neither condition above holds, then full connectivity cannot be assured. In such a case, the resulting metagraph has to be itself tested for full connectivity by looking for a metapath from its pure inputs to its pure outputs in each interpretation.

It is possible to determine whether a given metagraph is non-redundant in a given context, using the algebraic representation. The incidence matrix G can be adapted to the context by reducing it based on the elements in P (the true propositions) and Q (the false propositions). Then, the resulting metagraph is non-redundant if each row has at most one '$+1$' entry. Otherwise, there are two or more edges (tasks) that produce the same output. As before, the complexity of checking whether the metagraph is non-redundant in all contexts depends upon the number of valid contexts, which in turn is determined by R (the remaining undetermined propositions).

Chapter 6

INDEPENDENT SUB-METAGRAPHS

We now examine the issue of independence of a sub-metagraph contained within a larger metagraph. This is a useful notion, since it helps identify components of a larger and complex system that can be abstracted a a higher level, and possibly removed as a separate subsystem.

DEFINITION 6.1. A metagraph $S' = \langle X', E' \rangle$ is said to be a *sub-metagraph* (SMG) of another metagraph $S = \langle X, E \rangle$ (denoted by $S' \subseteq S$) if $X' \subseteq X$ and $E' \subseteq E$.

Note that the SMG relationship is defined by edges, not by elements. Thus, it is possible for $S' \subset S$ even if $X' = X$ as long as $E' \subset E$.

Input independence: A metagraph S_1 is an input independent SMG of a metagraph S_2 if every element of S_1 that is not a pure input is determined only by edges within S_1.

Output independence: A metagraph S_1 is an output independent SMG of a metagraph S_2 if every element of S_1 that is not a pure output is used (i.e. as an input) only by edges within S_1.

Independence: A metagraph S_1 is an independent SMG (denoted ISMG) of a metagraph S_2 if it is both input independent and output independent.

To see examples of input independence, output independence and independence consider Figure 6.1. A number of possible sub-metagraphs can be identified in this metagraph, and Table 6.1 lists a number of these, along with their pure inputs and outputs, and whether they are input independent, output independent and/or independent. From this example, it should be apparent that the independence properties of a particular SMG are not always readily apparent from visual inspection.

Table 6.1. Independence of some SMGs in Figure 6.1.

SMG	PI	PO	Other	Input Ind.	Output Ind.	Independent
$\{e_1, e_2\}$	a, h	d, k, l	c	NO	YES	NO
$\{e_1, e_2, e_3\}$	a, h	k, l	c, d	YES	NO	NO
$\{e_2, e_3, e_4, e_5\}$	c, d, h, q	l, n, p	k	YES	YES	YES
$\{e_1, e_2, e_4, e_5\}$	a, h, q	l, n, p	c, d, k	NO	NO	NO

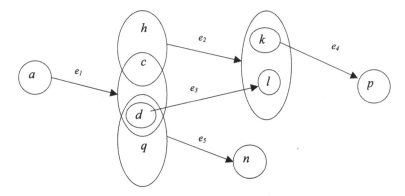

Figure 6.1. Illustration of independent components in a metagraph.

It is also useful to consider the relationships between different ISMGs of a given metagraph. The following theorems (from Basu and Blanning, 2003) identify some special cases of interest:

THEOREM 6.1. *Given two ISMGs S_1 and S_2 of a common containing metagraph S, then $S_3 = S_1 \cup S_2$ is an ISMG of S.*

PROOF. We prove this by contradiction. All elements that are not either pure inputs or pure outputs of S_1 or S_2 clearly cannot violate the independence of S_3. Let x be a pure input of S_1 that violates the independence of S_3 by being in the output of some edge outside S_3. Clearly x must be a pure output of S_2. But by definition, every pure output of S_2 is determined only by edges within S_2, and thus S_3, which contradicts the claim about x. A similar argument holds if x is a pure output of S_1 that is an input to some edge outside S_3. Thus the result is proved.   □

Independence is desirable in the sense that any coordination issues involving an ISMG can be addressed solely in terms of its pure inputs and pure outputs, while this is not true for all SMGs in general.

DEFINITION 6.2. Two metagraphs are *mutually independent* if each is an ISMG in the metagraph formed by their union.

THEOREM 6.2. *If two edge-disjoint SMGs (no common edges) are both ISMGs of a single containing metagraph, then they are mutually independent of each other.*

PROOF. Since the two SMGs are edge-disjoint and each is an ISMG, the only common elements can be "boundary elements" (i.e. pure inputs or pure outputs), and furthermore, the SMGs must be sequentially related, that is, PI_1 can overlap with PO_2 but not PI_2, and PO_1 can overlap with PI_2 but not PO_2 (the reason for this is that if PI_1 and PI_2 have any common elements, then because of the independence assumption, all edges containing the common elements have to be in both SMGs, which violates their edge-disjointedness). It then follows that they are each ISMGs of their union, which proves the result. $\quad\square$

THEOREM 6.3. *Given two ISMGs S_1 and S_2 of a containing metagraph S, the intersection $S_3 = S_1 \cap S_2$ is also an ISMG of S.*

PROOF. Each ISMG can be viewed as the union of a set of independent meta-paths (each itself an ISMG of S and S_1), each from some subset of pure inputs to a subset of pure outputs of that ISMG. By the definition of independence, if any edge e occurs in both S_1 and S_2, every metapath $M(PI_1, PI_2)$ in S_1 containing e must also occur in S_2. Otherwise, at least one edge in M would violate the independence of S_1 and S_2. Thus $S_1 \cap S_2$ can also be viewed as the union of a set of independent metapaths from its pure inputs to its pure outputs, and is thus an ISMG of S. $\quad\square$

Using the matrix representation of metagraphs, we have developed an algorithmic procedure to test for independence of a given SMG S' in a metagraph S. The procedure is quite straightforward, since the columns in A corresponding to all elements in S' other than the pure inputs have to be empty except for rows corresponding to elements in S'; similarly all rows corresponding to elements in S' other than pure outputs have to be empty except for columns corresponding to elements in S'.

The following algorithm determines whether a given sub-metagraph $S'(X', E')$ is independent within a given metagraph $S(X, E)$ with adjacency matrix A:

Procedure Check-Independence (S', A)

Let PI' (PO') be the set of pure inputs (pure outputs) of S' {generated using Procedure **PIPO** below}.

    For $i = 1, \ldots, |X|$
        For $j = 1, \ldots, |X|$
        If $[(x_i \in \{X' \backslash PI'\}$ and $x_j \notin X')$ or $(x_j \in \{X' \backslash PO'\}$ and $x_i \notin X')]$ and $a_{ij} \neq \varnothing$ then S' is not independent in S; STOP.
        Next j
    Next i
    S' is independent in S;
    STOP.

The following algorithm identifies the sets PI' and PO' used in procedure **Check-Independence**:

Procedure PIPO (S', A)

Let $PI' = PO' = \varnothing$;
For each $x_i \in X'$,
    If $a_{ij} = \varnothing \ \forall x_j \in X'$ then $PO' = PO' \cup x_i$;
    If $a_{ji} = \varnothing \ \forall x_j \in X'$ then $PI' = PI' \cup x_i$;
Repeat;
Return PI', PO';
STOP.

Note that these are polynomial procedures that are guaranteed to terminate.

PART II

APPLICATIONS OF METAGRAPHS

Chapter 7

METAGRAPHS IN MODEL MANAGEMENT

This is the first of three chapters in which we will examine three applications of metagraphs to information processing systems. The first is the application to the management of decision models, which is examined in this chapter. The second is the management of data bases and rule bases, which will be examined in Chapter 8. The third is the management of workflow systems, in which the work consists of information processing tasks to be performed by humans or machines. We will examine this application in Chapter 9.

There are four important topics in the application of metagraphs to model management. The first is the representation of decision models as metagraphs, in which the input-to-output mapping of a model is represented by the set-to-set mapping in a metagraph edge. Thus, a model base is a set of metagraph edges (i.e., a metagraph) which collectively represents the model base. We are not interested here in the content of the models; rather we view models as "black boxes" and consider only relationships among models as determined by common invertex and outvertex elements; for example, an output of one model may be an input to another model.

The second topic is model selection and integration. In both selection and integration the principal concept is that of a metapath from a set of known, or source, elements (i.e., elements whose values are assumed to be known) and a set of desired, or target elements (i.e., elements whose values are to be calculated). We are not concerned here with calculation procedures, but only with the existence and uniqueness of metapaths from source to target. In the case of multiple metapaths, we wish to identify any bridges – that is, edges that must be present in the metagraph regardless of which metapath is used. A distinction of interest here is the one between cyclic and acyclic metagraphs, which represent two quite different structures of model bases.

The third topic is hierarchical modeling, which concerns the integration of model bases with possibly overlapping variables – that is, the integration of metagraphs with possibly overlapping generating sets. Each model base may be used in a simplified analysis in which it is projected over a subset of its generating set. If the information in these projections is to be integrated, the question is whether the union of the two projections is the same as the projection of the union of the two metagraphs over the union of the two generating sets. In other words, can we combine (by taking the union) the two projections or must we first combine the two metagraphs and then do the projection. We will see that the union of the projections is not the same as the projection of the union. However, the projection of the union dominates the union of the projections.

The fourth topic is the use of assumptions in model bases. An assumption associated with a model is a proposition (i.e., a variable that can take on either of two values – true or false) that must be true for the model to be used in a calculation procedure (i.e., for an edge to be used in a metapath). Thus, a model base is viewed as a conditional metagraph in which the generating set is enlarged to include the assumptions as elements and an assumption associated with a particular model is part of the invertex of the edge representing the model. We consider the projection and context of a conditional metagraph and ask if they are commutative: given a conditional metagraph, a subset of the generating set, and a set of assumptions (with true, false, or unknown values), is the projection of the context the same as the context of the projection. We will see that they are the same.

1.   MODELS AS METAGRAPHS

In the metagraph view of models each model is represented as an edge with the inputs as invertex and the outputs as outvertex. The principal issue here is connectivity – that is, the existence or lack of existence of one or more metapaths connecting a source set of elements to a target set of elements. In the case where there is more than one metapath, we wish to know whether there are any bridges – that is, any edges in the intersection of all such metapaths. If there are any such edges, then they must exist for the source to be connected to the target. In other words, the corresponding models must exist if the source elements can be used to calculate the target elements.

Consider the simple example of a model base illustrated in Figure 7.1. There are four elements: inflation rate (*INFL*), which is a pure input, revenues realized (*REV*) and expense incurred (*EXP*), both of which depend on *INFL* and only on *INFL*, and the resulting net income (*NI*), which depends on both *REV* and *EXP*. Thus, the generating set is $X = \{INFL, REV, EXP, NI\}$. There are three models, which are represented by three edges. The first is a sales model, *sls*, which calculates its outvertex {*REV*} from its invertex {*INFL*}. The second is a cost model, *cost*, which calculates its outvertex {*EXP*} from its invertex {*INFL*}, and the third is a financial model, *fin*, which calculates its outvertex {*NI*} from its invertex {*REV, EXP*}.

We can see from the various components of the adjacency matrix A (Figure 7.2) that *sls* and *cost* have no coinputs or cooutputs, but that is not true of *rev*. For example, $a_{INFL,REV}$ discloses that *infl* needs only *INFL* as input (i.e., there are no coinputs) and produces only *INFL* as its output (there are no cooutputs). Similarly, $a_{INFL,EXP}$ discloses that *cost* needs only *INFL* as its

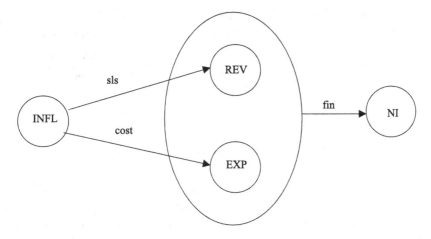

Figure 7.1. Simple model metagraph.

	INFL	REV	EXP	NI
INFL	∅	{<∅, ∅, \<sls\>>}	{<∅, ∅, \<cost\>>}	∅
REV	∅	∅	∅	{<EXP, ∅, \<fin\>>}
EXP	∅	∅	∅	{<REV, ∅, \<fin\>>}
NI	∅	∅	∅	∅

Figure 7.2. Adjacency matrix for Figure 7.1.

input and produces only *EXP* as its output. On the other hand, the *NI* column of *A* discloses that *fin* has both *EXP* and *REV* as inputs but only *NI* as output. That is, $a_{REV,NI}$ has *EXP* as coinput and $a_{EXP,NI}$ has *REV* as coinput, but in both cases *NI* has no cooutput.

The closure of the adjacency matrix A^*, is illustrated in Figure 7.3, shows that there are two simple paths of length exceeding 1, both connecting *INFL* to *NI*. The first is the simple path ⟨*sls*, *fin*⟩, which has *EXP* as coinput and *REV* as cooutput. The second is the simple path ⟨*cost*, *fin*⟩, which has *REV* as coinput and *EXP* as cooutput. Thus, there is no sequence of models (i.e., no simple path) connecting *INFL* to *NI* that is free of coinputs. However, this does not mean that *NI* cannot be calculated from *INFL* alone, for there is a meta-path {*sls*, *cost*, *fin*} connecting *INFL* to *NI*. Thus, we can see the advantage of representing model bases as metagraphs and of defining metapaths: metapaths disclose a type of connectivity that simple paths do not.

Finally, we can see that the model bank is acyclic, since A^* contains only null diagonal elements. Thus, there is no sequence of edges (simple path) from any element to itself, and it is possible to execute the models in a sequential

	INFL	REV	EXP	NI
INFL	∅	{<∅, ∅, <sls>>}	{<∅, ∅, <cost>>}	{<{EXP}, {REV}, <sls, fin>>, <{REV}, {EXP}, <cost, fin>>}
REV	∅	∅	∅	{<EXP, ∅, <fin>>}
EXP	∅	∅	∅	{<REV, ∅, <fin>>}
NI	∅	∅	∅	∅

Figure 7.3. Closure A^* of the adjacency matrix for Figure 7.1.

fashion without the need for iterative execution or other simultaneous implementation of the models. This is accomplished by executing *rev* and *exp* (in either order) and then *fin*. Of course, if there were an edge connecting *NI* to *INFL* we would have a cyclic model bank and it would be necessary of resolve the simultaneity. But it is unlikely that such a model would exist, since the inflation rate in the economy would presumably be exogenous to a particular firm. Thus, the model bank and its metagraph are acyclic, and the models (edges) form a partially ordered set.

2.   MODEL SELECTION AND INTEGRATION

This is not the case with the model base illustrated in Figure 7.4, which describes the supply and demand relationships in a firm and its market. There are four elements: the gross national product or other measure of overall economic activity, *GNP*, the price charged by the firm, *PRI*, the resulting sales volume, *VOL*, and the firm's capacity to produce, *CAP*. There are two models, a demand model, *dmd*, with invertex {*GNP, PRI*} and outvertex {*VOL*} and a supply model, *sup*, with invertex {*VOL*} and outvertex {*PRI, CAP*}. This is a cyclic metagraph describing a cyclic model base.

We will not describe the entire adjacency matrix, but we note that it contains two non-null diagonal elements:

$$a^*_{PRI,PRI} = \big\{ \{GNP\}, \{CAP, VOL\}, \langle dmd, sup \rangle \big\},$$
$$a^*_{VOL,VOL} = \big\{ \{GNP\}, \{CAP, PRI\}, \langle sup, dmd \rangle \big\}.$$

Thus, there is a simple path, $\langle dmd, sup \rangle$, from *PRI* to itself and another simple path, $\langle sup, dmd \rangle$, from *VOL* to itself. Since these two simple paths form a cyclic permutation of each other, there is a single cycle involving both *PRI* and *VOL*. We note that $a^*_{GNP,GNP}$ and $a^*_{CAP,CAP}$ are both null, since *GNP* and *CAP* do not participate in any cycles.

In this case the models do not form a partially ordered set, because we need to know the value of *PRI* to calculate *VOL* and we need to know the value of

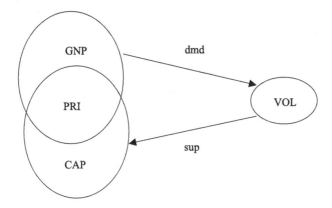

Figure 7.4. Cyclic metagraph.

VOL to calculate *PRI*. If we were able to look inside the black boxes and we were to find that the models had a simple functional form, such as linearity, we could determine equilibrium values for *PRI* and *VOL* as a function of *GNP* (which would presumably be a parameter in the ⟨{*GNP*, *PRI*}, {*VOL*}⟩ relationship) in closed form. But since we assume that the models are atomic (i.e., they are black boxes), that is not possible and an iterative approach is needed.

The iterative approach would begin by receiving an exogenous input *GNP*. Then there are two possible sequences of activities:

- We would posit an initial value for *PRI*, enter these values for *GNP* and *PRI* into the model *dmd* to calculate *VOL*, and then use *sup* to calculate a second value for *PRI*. The initial and calculated values for *PRI* would be compared. If they were sufficiently close to each other (within a predetermined threshold), the process would terminate. Otherwise, a new initial value for *PRI* would be calculated, possible by splitting the difference between the posited and calculated values of *PRI*, and the calculation cycle would begin anew. This would continue until the process converged.
- Alternatively, we could begin with an exogenous value of *GNP*, as before, but posit an initial value of *VOL*. Then *sup* would be used to calculate a value for *PRI*, which would be entered into *dmd* to calculate a second value for *VOL*. If the initial and calculated values were sufficiently close to each other, the process would terminate, otherwise the cycle would be repeated.

We note that in each of these cases a value for *CAP* would be calculated during each cycle, but only the final value would be of interest.

In describing the iterative approach we have relied on the notion that equilibrium values for *PRI* and *VOL* exist and that this equilibrium is unique (i.e., that there are not multiple values of *PRI*, *VOL* pairs that are within the pre-

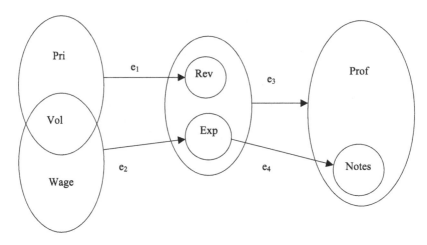

Figure 7.5. Metagraph with bridges.

determined thresholds). If the first notion is false, we would have to terminate the search once the lack of an equilibrium is apparent. In the second case we would have to identify the various equilibria and select an appropriate one.

We now turn to a second topic, multiple simple paths connecting two elements. This may lead to multiple metapaths between the same source and target. Consider the example of Figure 7.5, in which e_1, e_2, and e_3 are bridges between their respective elements. For example, it would be impossible to calculate *Rev* from *Pri* and *Vol* without e_1. However, e_4 is not a bridge between {*Rev*, *Exp*} and {*Notes*} because either e_3 or e_4 could be used to perform the calculation. This can be seen from $a_{EXP,NOTES}$, which consists of two components $\langle \{Rev\}, \{Prof\}, \langle e_3 \rangle \rangle$ and $\langle \varnothing, \varnothing, \langle e_4 \rangle \rangle$.

3.   HIERARCHICAL MODELING

To illustrate hierarchical modeling, we use an example from life cycle costing. The example, in metagraph form, appears in Figure 7.6, and the legend defining the elements (variables) in the generating set are defined in Figure 7.7. For example, model e_1 will calculate the manufacturing cost per vehicle (*MC*) and the expected service life (*SL*) from the design variables (*DV*), such as the engine volume and the assembly time.

We now consider the projection of the metagraph in Figure 7.6 over a subset of its generating set. Let $X' = \{DV, MD, PR, LO, TL\}$. The projection, which appears in Figure 7.8, consists of three edges, each of which is a dominant metapath in *S*, and the vertices of which are contained in X'. No elements in $X \backslash X' = \{MC, SL, AO\}$ appear in the projection.

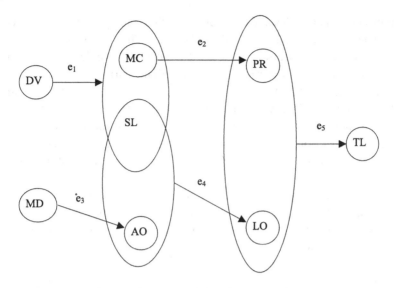

Figure 7.6. Metagraph for life cycle costing.

AN	Annual fuel consumption	AO	Annual operating & support cost
AS	Annual fuel cost	DV	Design variables
FC	Fuel cost	FQ	Frequency of maintenance
LO	Life cycle ops & support cost	MC	Manufacturing cost
MD	Annual miles driven	PM	Annual cost of meintenance
PR	Price	SL	Expected service life
SN	Sensitivity of op. cost to MD	TL	Total life cycle cost per vehicle

Figure 7.7. Variables used in Figure 7.6.

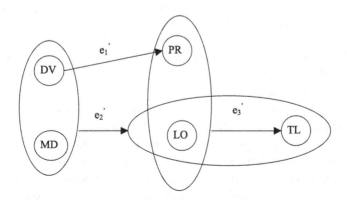

Figure 7.8. Projection of life cycle costing metagraph.

The projection provides a high level view of the metagraph that hides certain details. The projection will be incomplete, and deliberately so – that is, if an edge $e' = \langle V', W' \rangle$ appears in S', then it is possible to calculate W' from V' in S, but there may be several other intermediate variables in $X \backslash X'$ that are also calculated. For example, in Figure 7.8 we can see that it is possible to calculate the price (PR) given only the design variables (DV), and the fact that manufacturing cost (MC) is an intermediate variable is hidden from the person viewing the projection, since $MC \in X \backslash X'$. In addition, the fact that service life (SL) is also calculated in the process (by e_1) is hidden from the user, because $SL \in X \backslash X'$.

Each of the edges in the projection may be of interest to one or more managers or analysts in an organization. For example, e'_1 may be of interest to a marketing manager who wishes to know how the design of a vehicle will affect its price, e'_2 may be of interest to the manager of a car rental company who wants to evaluate different designs under various mileage conditions, and e'_3 may be of interest to senior managers who wish to know the profit consequences of various pricing strategies. Thus, a projection may have several "customers" who see different benefits to the different edges. Of course, this can also be accomplished by making several projections of a single metagraph.

The advantage of a projection is that it may disclose relationships that are implicit in the original metagraph but are not easy to see because of the size and complexity of the original metagraph. The relationship, represented by e'_1 between DV and FR is one example. Another example is provided by e'_2, which represents the invocation of a metapath $M(\{DV, MD\}, \{LO, TL\}) = \{e_1, e_2, e_3, e_4, e_5\}$. It may not be immediately clear that this is a dominant metapath for the calculation of both LO and TL. The third edge in S', e'_3 is easily discernible from S, since it is simply e_5.

Thus, in Figures 7.6 and 7.8, $C(e'_1) = \{\{e_1, e_2\}\}$, $C(e'_2) = \{\{e_1, e_2, e_3, e_4, e_5\}\}$ and $C(e'_3) = \{\{e_5\}\}$. We note that a composition is not a set of edges, but a set of sets of edges, because there may be more than one metapath in S corresponding to an edge in S'. The composition of edges in this and all other projections in this chapter are shown in Figure 7.9. In the simple examples presented here, there are no alternate metapaths between any source and target; therefore, each composition in Figure 7.7 contains a single set of edges.

Thus, a set of models such as the life cycle costing models can be transformed into a higher-level view in which certain variables and relationships are retained and others are deliberately hidden from the user. In the example given above, several managers wished to know about relationships between the design variables, miles driven, sales price, annual operating costs, and total life cycle costs. On the other hand, they were not interested in the manufacturing costs, service life, or the sum of annual operating and service costs. Where these latter variables appeared as inputs to a model, the model was discarded

$C(e_1{}')$ (Fig 7.8)	$\{\{e_1, e_2\}\}$	$C(e_2{}')$ (Fig 7.8)	$\{\{e_1, e_2, e_3, e_4, e_5\}\}$
$C(e_3{}')$ (Fig 7.8)	$\{\{e_5\}\}$	$C(e_4{}')$ (Fig 7.11)	$\{\{e_6, e_9\}\}$
$C(e_5{}')$ (Fig 7.11)	$\{\{e_7\}\}$	$C(e_6{}')$ (Fig 7.13)	$\{\{e_1, e_2, e_3, e_4, e_5\}\}$
$C(e_7{}')$ (Fig 7.13)	$\{\{e_1, e_2\}\}$	$C(e_8{}')$ (Fig 7.13)	$\{\{e_3, e_7\}\}$
$C(e_9{}')$ (Fig 7.13)	$\{\{e_4\}\}$	$C(e_{10}{}')$ (Fig 7.13)	$\{\{e_5\}\}$
$C(e_{11}{}')$ (Fig 7.15)	$\{\{e_1\}\}$	$C(e_{12}{}')$ (Fig 7.15)	$\{\{e_3\}\}$
$C(e_{13}{}')$ (Fig 7.15)	$\{\{e_1, e_3, e_4\}\}$	$C(e_{14}{}')$ (Fig 7.15)	$\{\{e_4\}\}$
$C(e_{15}{}')$ (Fig 7.16)	$\{\{e_1\}\}$	$C(e_{16}{}')$ (Fig 7.16)	$\{\{e_3, e_4, e_7\}\}$

Figure 7.9. Compositions of projected edges in various projections.

in the higher-level view. Where they only appeared as outputs, the models were retained, but the outputs were restated to exclude these variables. Where the discarded variables were intermediate outputs in relationships between variables of interest to the managers, the relationships appeared in the higher-level views, without the intermediate variables. For example, a marketing manager could see that the design variables were sufficient to determine the sale price of the car, without having to consider, or be even aware of the intervening variable manufacturing cost.

In addition, several views may be constructed from a single model base. In other words, different managers accessing the same model base may have different views. For example, a marketing manager negotiating with a major customer may be interested only in the design variables, the annual operating and service costs, and the total life cycle cost. If we construct a higher-level view of the life cycle models with only these variables, we find a single relationship: that the design variables and the annual operating and service costs are sufficient to compute total life cycle costs. All the other variables are deliberately hidden from this view.

We have examined the use of metagraphs in constructing views of a single model base. We now turn to the issue of combining model bases. Consider a situation where two sets of users have distinct model bases over possibly overlapping generating sets. Also, consider two users, one in each group, that visualize their respective model bases in terms of specific projections. For instance, a marketing manager may use a view that relates demand and manufacturing cost to price and volume, while a production manager may use a view that contains a relationship between batch size and raw, material cost to unit manufacturing cost and mean time between failures. The two managers may want to combine their resources to solve aggregate problems, such as determining the effect of change of batch size on sales volume. The underlying analytical question that this raises is whether it is sufficient to simply combine the views that the two managers have of their respective model bases, or whether it is necessary to combine the model bases and then project them.

The conceptual problem underlying this issue is the combination of two metagraphs to produce a new metagraph. If the metagraphs are $S_1 = \langle X_1, E_1 \rangle$ and $S_2 = \langle X_2, E_2 \rangle$, then the new metagraph, which we will call the sum of S_1 and S_2, will be $S_{12} = S_1 + S_2 = \langle X_1 \cup X_2, E_1 \cup E_2 \rangle$. If $X_1 \cap X_2 \neq \emptyset$, then S_{12} may contain simple paths and metapaths that are not in either S_1 or S_2. To see this, consider two projections S_1' and S_2' of S_1 and S_2 respectively, where $S_1' = \langle X_1', E_1' \rangle$ is the projection of S_1 over $X_1' \subseteq X_1$ and $S_2' = \langle X_2', E_2' \rangle$ is the projection of S_2 over $X_2' \subseteq X_2$. If we combine the two views, we get a metagraph $S_1' + S_2' = \langle X_1' \cup X_2', E_1' \cup E_2' \rangle$. We would like to know whether any information about relationships between elements of $X_1' \cup X_2'$ are lost in this process. That is, we would like to know whether $S_1' + S_2'$ contains the same information as the metagraph $S_{12}' = \langle X_1' \cup X_2', E_{12}' \rangle$, which is the projection of S_{12} over $X_1' \cup X_2'$.

We note that S_{12}' dominates the sum $S_1' + S_2'$, but the converse need not hold. That is, there may be some edges in S_{12} that are not dominated by any edges in $S_1' + S_2'$. To illustrate this, we expand the life cycle costing example described in the previous section. Consider a set of four cost estimating relationships (variables are defined in Figure 7.7). The metagraph is illustrated in Figure 7.10. The four edges $e_6, \ldots, e_9$ are cost estimating relationships that would be constructed by engineers and used by the sales force. Figure 7.11 contains the projection of this metagraph over $\{FC, PM, MD, SL, AO\}$. The edge e_4' allows us to calculate AO from FC, PM and MD without consideration of the intervening variable, AS, and $C(e_4') = \{\{e_5, e_9\}\}$. The calculation represented by e_5' is simply that represented by e_7, so $C(e_5') = \{\{e_7\}\}$.

We will denote the metagraph in Figure 7.6 as S_1, Figure 7.8 as S_1', Figure 7.10 as S_2, and Figure 7.11 as S_2'. Also $X_1' = \{DV, MD, PR, LO, TL\}$ and $X_2' = \{FC, PM, MD, SL, AO\}$; thus, $X_1' \cup X_2' = \{DV, MD, PR, LO, TL, FC, PM, SL, AO\}$. The joint metagraph – that is, the sum of S_1 and S_2 – appears in Figure 7.12, and its projection over $X_1' \cup X_2'$, giving S_{12}', appears in Figure 7.13.

The sum of the projections, $S_1' + S_2'$, appears in Figure 7.14. We can see from Figures 7.13 and 7.14 that S_{12}' dominates $S_1' + S_2'$. For example, some edges, such as $\langle \{PR, LO\}, \{TL\} \rangle$, appear in both S_{12}' and $S_1' + S_2'$. It appears as e_{10}' in S_{12}' and e_3' in $S_1' + S_2'$. In this case the compositions of the relevant edges are the same: $C(e_1') = C(e_3') = \{\{e_5\}\}$. On the other hand, the edge $e_4' = \langle \{MD, FC, PM\}, \{AO\} \rangle$ in $S_1' + S_2'$ does not appear in S_{12}', but it is dominated by the edge $e_8' = \langle \{MD\}, \{AO, SL\} \rangle$ in S_{12}'. In addition, there are edges in S_{12}' that

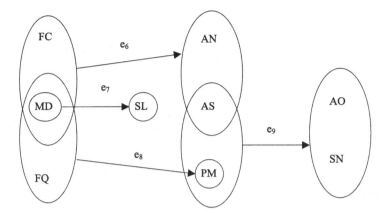

Figure 7.10. Metagraph of cost estimating relationships.

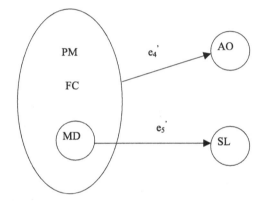

Figure 7.11. Projection of cost estimating metagraph.

do not dominate any edge in $S_1' + S_2'$. An example is $e_1' = \langle \{DV\}, \{PR, SL\} \rangle$. Although the edge $e_1' = \langle \{DV\}, \{PR\} \rangle$ is found in $S_1' + S_2'$, there is no edge in $S_1' + S_2'$ with DV in its invertex and SL in its outvertex.

A possible negative consequence of integrating high-level views of a model base is that information may be lost. In the metagraph $S_1' + S_2'$, we find that mileage driven, along with fuel cost and the annual cost of preventive maintenance, is sufficient to determine annual operation and support cost. However, we do not find that mileage driven by itself is sufficient to determine annual operation and maintenance cost, because the latter variable is not in the set of elements over which the life cycle costing metagraph was projected. In addition, the fact that the design variables are sufficient to determine expected service life is missing from the sum of the projections, even though both variables are found in the sum of the projection, because expected service life was

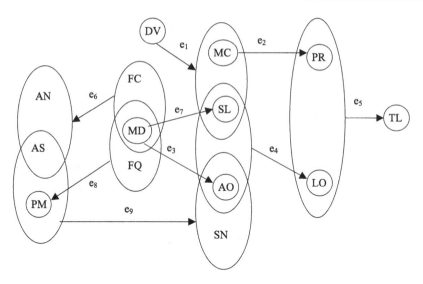

Figure 7.12. The joint metagraph.

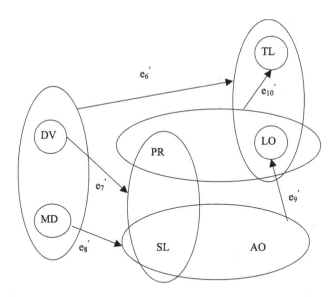

Figure 7.13. Projection of the joint metagraph.

not in the set of elements over which the life cycle costing metagraph was projected. This may give the misleading impression that service life can be calculated only from miles driven, even though it can also be calculated from the design variables.

The condition $X_1 \cap X_2 = X_1' \cap X_2'$, did not hold in the previous example, because $X_1 \cap X_2 = \{MD, SL, AO\}$ but $X_1' \cap X_2' = \{MD\}$. Thus there were el-

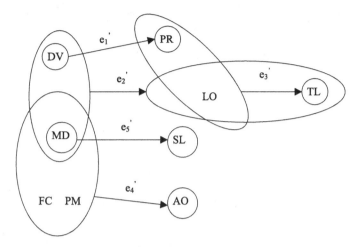

Figure 7.14. The sum of the projections (Figures 7.8 and 7.11).

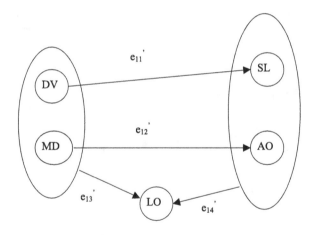

Figure 7.15. Second projection of Figure 7.6.

ements (*SL* and *AO*) common to the generating sets of metagraphs that were not common to the sets of elements over which the two metagraphs were projected.

We now construct an example in which the condition is satisfied. We will retain the projection of S_2 over $\{FC, PM, MD, SL, AO\}$ (in Figure 7.11) but project S_1 over a new $X'_1 = \{DV, SL, MD, AO, LO\}$ (in Figure 7.15). In this case a new $X'_1 = X_1 \cap X_2 = X'_1 \cap X'_2 = \{MD, SL, AO\}$ – that is, all of the elements that are found in the generating sets of both S_1 and S_2 are also in the sets of elements over which S_1 and S_2 are projected. The union of the sets over which the metagraphs are projected is $X'_1 \cup X'_2 = \{DV, SL, MD, AO, LO, FC, P\}$, and the projection of S_{12} over this set, which

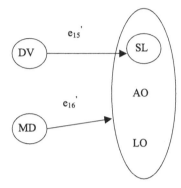

Figure 7.16. Second projection of the joint metagraph.

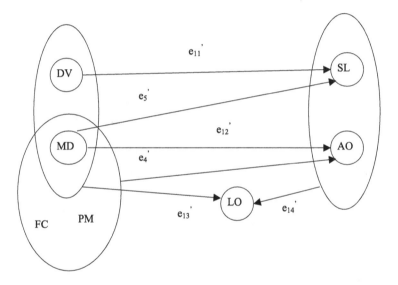

Figure 7.17. The sum of Figures 7.11 and 7.15.

is a new S'_{12}, is shown in Figure 7.16. The sum of the projections, $S'_1 + S'_2$, appears in Figure 7.17.

We can see as before that S_{12} dominates $S'_1 + S'_2$. For example the edge $\langle \{DV\}, \{SL\} \rangle$ is present in both S_{12} and $S'_1 + S'_2$, whereas in the previous example it was not present in $S'_1 + S'_2$. The edge $e_4 = \langle \{MD, FC, PM\}, \{AO\} \rangle$ is in S'_{12}. We can also see that the converse in true; $S'_1 + S'_2$ dominates S'_{12}. For example, the edge e'_{16} in S'_{12} is dominated by the metapath $\{e'_5, e'_{12}, e'_{14}\}$ in $S'_1 + S'_2$.

We can also see that equivalence (i.e., mutual dominance) does not necessarily imply equality. The metagraphs in Figures 7.16 and 7.17 are not the same, in part because of the existence of edge $e_4 = \langle \{MD, FC, PM\}, \{AO\} \rangle$

in $S_1' + S_2'$. However, this edge does not destroy the equivalence of S_{12}' and $S_1' + S_2'$, since it is dominated by e_{16}' in $S_1' + S_2'$. In addition, the edge $e_1' = \langle\{DV, MD\}, \{LO\}\rangle$ in $S_1' + S_2'$ does not appear in S_{12}', but it is dominated by the edge S_{12}', which in turn does not appear in $S_1' + S_2'$ but is dominated by the metapath $\{e_5', e_{12}', e_{13}'\}$ in $S_1' + S_2'$.

Although S_{12}' and $S_1' + S_2'$ are equivalent, there is a difference between them: S_{12}' seems simpler than $S_1' + S_2'$. One reason for this is that a projection such as S_{12}' only contains dominant metapaths, whereas the sum of two projections such as $S_1' + S_2'$, need not. For example, in Figure 7.17, the edge $e_4' = \langle\{MD, FC, PM\}, \{AO\}\rangle$ is in $S_1' + S_2'$, but not in S_{12}', because it is dominated by e_{12}'. Another reason for the simplicity of S_{12}' is the requirement that no two edges can have the same invertex. Thus, edge e_{16}' in S_{12}' corresponds to three edges – e_5', e_{12}', and e_{14}' – in $S_1' + S_2'$. As before, there is no such simplifying requirement for the sum of two projections, only for a single projection.

We have seen that there is a simple criterion for the integration of two views that avoids the misleading impression that a calculation cannot be performed (e.g., that expected service life cannot be calculated from the design variables) when in fact the calculation can be performed. The requirement is that all variables common to the two sets of calculations (i.e., the life cycle costing calculations and the cost estimating relationships) also be in both of the sets of variables used to construct the higher-level views. We have also seen that two views can be equivalent without being identical. In our second example, the sum of the higher-level views of the life cycle costing calculations and of the cost estimating relationships was equivalent to a single view constructed directly from both sets of calculations.

At the same time, the ways in which these relationships were presented to the user were different in two respects. First, the sum of the high-level views contained redundant information, in the form of dominated metapaths. For example, the sum of the views disclosed that miles driven, fuel cost, and annual cost of preventive maintenance are sufficient to calculate annual operation and support cost, but it also disclosed that miles driven is sufficient to calculate annual operation and support cost. The second difference is that several relationships in the sum of the higher-level views may appear as only one relationship in the direct view of both sets of calculations. For example, the sum of views contained separate relationships between miles driven and service life, miles driven and annual operations and service cost, and miles driven and life cycle operation and support cost. In the direct view this is presented to the users as a single relationship: miles driven is sufficient to calculate all three variables. Thus, the users can easily see that several of the variables of interest to them are determined by a single variable, miles driven.

4.   ASSUMPTIONS IN MODEL BASES

We now turn to the use of assumptions in model bases. We define an assumption as a proposition (i.e., a statement that may be true or false) associated with a model that must be true if the model is to be used in a particular instance. In metagraph terms, an assumption associated with an edge must be true for the edge to appear in a metapath.

In order to explore this we must use a conditional metagraph, in which the generating set is partitioned into two subsets. The first is a set of variables, denoted X_v, and the second is a set of propositional statements, denoted X_p. Each $x \in X_v$ represents a variable, such as revenue, production level or inflation rate. Each $p \in X_p$ represents a proposition, such as "The inflation rate is five percent or less", or "$INFL \leq 0.05$".

A variable that appears in the invertex of an edge represents an input to the model represented by the edge. However, a propositional statement that appears in the invertex of an edge does not represent an input but rather an assumption that must be true for the model to be valid. For instance, $p =$ "$INFL \leq 0.05$" is in the invertex of an edge e representing a model, then the model only applies when the inflation rate is less than 5%. When a proposition appears in the outvertex of an edge, the edge represents a procedure for determining whether the proposition is true or false. For example, an edge $\langle\{PRI\}, \{p\}\rangle$ where $p =$ "$PRI \leq 10$" represents a procedure for evaluating the proposition "$PRI \leq 10$" from the value of PRI.

Conditional metagraphs must meet two constraints in addition to the requirement that the generating set be partitioned into variables and propositions. First, for each edge, at least one of the vertices must be nonempty, and the invertex and outvertex of each edge must be disjoint. Second, if an outvertex contains a proposition, then it cannot contain any other element. Even with these constraints, a conditional metagraph is a generalization of the type of metagraph discussed above. That is, these previously defined metagraphs can be viewed as conditional metagraphs in which $X_p = \emptyset$.

The constraints on conditional metagraphs are illustrated in Figure 7.18, which represents a price–volume relationship. In Figure 7.18 (parts (a), (b), (c)) the assumption p depends on an input variable, two variables not in the input (i.e., inflation rate and a price index), and on no variable at all, respectively. In the last case (Figure 7.18(c)), which is called a pure input (as it depends on no other elements), the user would be asked whether the proposition p is true. In Figure 7.18(d) the validity of the model depends on the volume which is not allowed, since volume is an output of the model. Finally, the edge in Figure 7.18(e) is invalid because two propositions occupy a single outvertex.

The use of propositions to represent assumptions is reasonable, since it allows the representation of any assumption that can be stated as a declarative

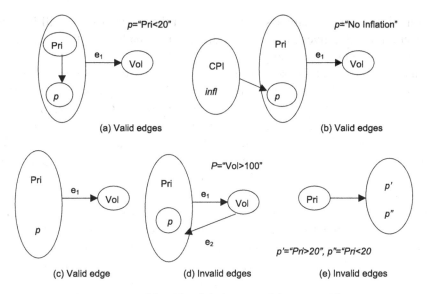

Figure 7.18. Valid and invalid edges containing propositions.

sentence, and this is true of most assumptions in modeling. The simplicity of the propositional representation, facilitates the inclusion of assumptions in a metagraph representation of a model base and allows a variety of important model management issues to be addressed: also, the use of metagraph edges to, evaluate propositions is quite general. This black box representation implies that the procedure to evaluate the proposition can have any structure, such as a simple calculation; an estimation model, or even a rule (in the case of an edge whose invertex consists of only propositions). While metagraph edges can be used to represent rules and rule based inference, modeling inference is outside the scope of this chapter. Thus, given an edge with a proposition as its outvertex, we simply use this to mean that given (appropriate) values of the invertex elements, the outvertex proposition can be evaluated.

A conditional metapath establishes a relationship between two sets of variables, using whatever assumptions are necessary to execute the necessary edges. The relevant assumptions are those that appear in the metapath. The propositions for each of the assumptions appear in the invertex of at least one of the edges in the conditional metapath and it may appear in one of the outvertices as well. If a proposition does not appear in any of the outvertices, it is a member of the set of initial assumptions; otherwise, it is a member of the set of intermediate assumptions.

The set of initial assumptions can be evaluated before any of the edges in the conditional metapath are executed. The intermediate assumptions are evaluated based on some of the outvertex elements of edges in the metapath, once the values of those elements are known. The truth falsity of any, other assumptions – that is, any assumptions not in the relevant set – will have no impact on the effectiveness of the conditional metapath in linking the source elements to the target elements.

In order to illustrate these concepts, we present a simple example: consider the metagraph in Figure 7.19 and the conditional metagraph, in Figure 7.20. In this example, the variables are $X_v = \{ADV, CAP, CC, ECON, EQT, EXP, NI, PRI, REV, STK, UCOST, VOL\}$ (these variables are defined in the caption of Figure 7.19). The propositions are $X_p = \{adl, cadv, inf, mkt, pkv, vdsk\}$, where adl means "advertising expense per unit is less than 25 percent of unit cost", $cadv$ means "competitive advertising does not increase more than 20 percent", inf means "inflation is less than 10 percent", mkt means "market conditions are stable", pkv means "peak volume does not exceed 3 million units", and $vdsk$ means "there are no volume discounts". The set of edges is $E = \{e_1, e_2, \ldots, e_9\}$, where e_1 is a pricing model for computing price whenever inflation is less than 10 percent, e_2 is a sales forecasting model that calculates sales when competitive advertising increases by no more than 20 percent, e_3 computes is a revenue forecasting model when peak volume is: no more than 3 million units and no volume discounts apply; e_4 is an accounting model that calculates total expense when advertising expenses are less than 25 percent of unit cost, e_5 computes net income, e_6 is a financial model computing cost of capital and stock price under stable market conditions, e_7 calculates unit cost, e_8 determines whether peak volume exceeds 3 million units, and e_9 determines whether adv is true.

If we consider the conditional metapath $M(\{ADV, CAP, ECON, UCOST\}, \{NI\}) = \{e_1, e_2, e_3, e_4, e_5, e_8, e_9\}$ from $\{ADV, CAP, ECON, UCOST\}$ to $\{NI\}$, then we have $\alpha = \{adl, cadv, inf, pkv, vdsk\}$, $\beta = \{cadv, inf, vdsk\}$, and $\gamma = \{adl, pkv\}$. That is, all of the assumptions except for mkt must be true for the metapath to represent a valid integrated model. However, only pkv and adl are intermediate assumptions that must be evaluated using e_8 and e_9; the others do not depend on any variables calculated during the execution of the edges in the metapath, and therefore can be evaluated at the outset.

Note that the interpretation of a metapath as a specification of an integrated model with the source elements as inputs and the target elements as outputs is slightly modified in the case of a conditional metagraph. In a simple metagraph (where $X_p = \varnothing$), given the values of the source elements, the integrated model represented by a metapath $M(B, C)$ unconditionally computes values for the target elements. However, in a conditional metagraph, just as the validity of a single edge is conditional upon the satisfiability of its assumptions, similarly, the validity of a conditional metapath as an integrated model is conditional upon the satisfiability of all the applicable assumptions in its edges.

When multiple assumptions are relevant to a problem instance, one question that arises is whether the assumptions are mutually consistent. For example, if

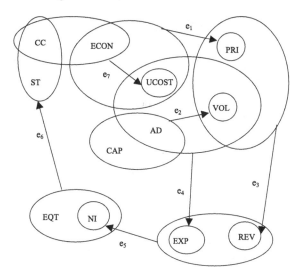

Figure 7.19. Model base metagraph: *ADV* – annual ad level; *CAP* – prod. capacity; *CC* – cost of capital; *ECON* – econ. indicator; *EQT* – total equity; *EXP* – total expense; *NI* – net income; *PRI* – sales price; *REV* – annual revenue; *STK* – stock price; *UCOST* – unit cost; *VOL* – vector of monthly sales.

two assumptions occurring in a metapath are p_1: "the inflation rate is less than 7 percent" and p_2: "the inflation rate is at least 10 percent", then clearly there is no feasible context in which both of these assumptions could be true. There is a substantial literature on consistency and integrity checking in knowledge bases (e.g., de Kleer, 1986, Grant and Minker, 1991, Illarramendi, Blanco and Goñi, 1994, and relatively simple procedures for consistency checking in propositional knowledge bases. However, since we do not use negated propositions in our approach, the knowledge base is implicitly consistent. Thus, recognition that p_1 and p_2 above are inconsistent would require manual reasoning. Toward this end, detection of the potential "conflict set", of propositions in a metapath can be done by identifying the set $(\bigcup_{e \in M(B,C)} V_e) \cap X_p$.

We also introduce the concept of a critical assumption. If there are two or more conditional metapaths from a given source to a given target, then the edges to be executed depend upon the particular metapath selected. In addition, the assumptions to be invoked may also depend upon the choice of metapath. Thus, the user may consider two factors in deciding how to calculate the target elements. The first is the set of models (edges) to be used, and the second is the set of applicable assumptions (and possibly even the issue of which assumptions are initial and which are intermediate). We define the set of critical assumptions for a given source and target as the intersection of the sets of relevant assumptions for all metapaths connecting the source to the target.

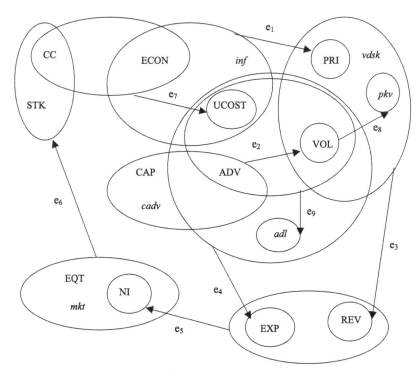

Figure 7.20. Conditional metagraph for model base in Figure 7.7: *adl – ADV* < 25% of *UCOST*; *inf* – inflation < 10%; *pkv* – peak volume < 3 MM; *cadv* – competitor's ad exp. increase < 20%; *mkt* – market conditions stable; *vdsk* – no volume discounts.

Thus, the critical assumptions are those that must hold if the available model base is to be used to compute the target elements given the source elements.

So far, we have discussed the use of metagraphs to support manipulation of models in a "flat" collection of models. However, in situations where the model base is quite large, it is useful to extract relevant views of the model base. In the metagraph representation of a model base, such views can be defined in two ways, as described in this section. We first discuss the concept of a context, and the use of assumptions in metagraphs to define contexts for problem solving. We then examine the role of projections, which provide a second mechanism for defining specialized views of a metagraph, and the use of projections in conditional metagraphs. Then we present a useful result about the relationship between these two types of views.

To define a context in a conditional metagraph, one begins by partitioning the assumptions in the metagraph into three sets: those known to be true, those known to be false, and those whose truth values are unknown. Then, the conditional metagraph is simplified so that only the last set of assumptions is present.

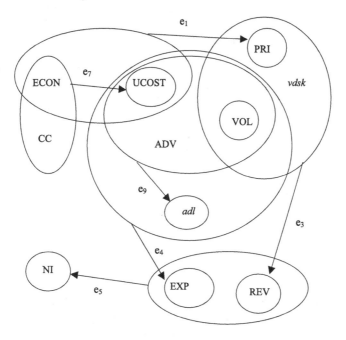

Figure 7.21. Context $K(\{inf, pkv\}, \{cadv, mkt\}, S)$.

In transforming a conditional metagraph into a context, the assumptions whose truth values are known (i.e., $P \cup Q$) no longer appear and need not be considered in the model selection process. Thus, a context represents a simplified view of a model base that allows a user to consider only those models known to be relevant, and those variables and assumptions whose values can be manipulated (e.g., in a sensitivity analysis). The larger the sets P and Q (i.e., the more specific the context), the simpler is the resulting conditional metagraph.

Consider again the conditional metagraph in Figure 7.19. If we know that propositions *inf* and *pkv* are true, and that *cadv* and *mkt* are false, then the resulting context $K(\{inf, pkv\}, \{cadv, mkt\}, S)$ is the conditional metagraph shown in Figure 7.21. Since *vdsk* and *adl* are the only propositions whose truth values are not specified in the context, they are the only assumptions appearing in Figure 7.21.

Note that elimination of models from a context does not mean that these models are removed from the model base, since a context is only a view. Rather, these models are not considered for problems defined within that context.

Model management systems, like data management systems, are often used by several people or groups, each of whom has a different purpose in mind. Thus it may be convenient, or even necessary, to present each user or user

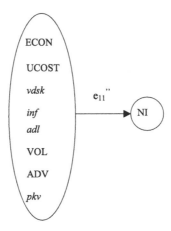

Figure 7.22. $N(\{ADV, ECON, UCOST, VOL\}, S)$.

group with a specialized view of the model base. This has two advantages. The first is convenience: the users are not burdened with information about models that they do not need to arrive at decisions or judgements that concern them. The second is security: if certain users should not be granted access to certain models, it is better that they do not even know that those models exist.

We note two interesting features of conditional metagraphs. First, the variables in the projection are limited to those in X', and any variables in $X \backslash X'$ will not appear. Some or all of the assumptions in X_p may appear in the projection; the user only specifies X', and the necessary assumptions are determined by the definition. Second, the projection (and therefore its adjacency matrix) represents all relationships among the variables in X' and $\alpha_{ij}'^* = \varnothing$ iff $\alpha_{ij}' = \varnothing$ (where A' is the adjacency matrix of the projection).

The first observation implies that the decision maker needs only to specify the relevant variables over which the projection is desired, and the operation then generates the relevant assumptions for each projected relationship. The second observation implies that use of projections does not require the computation of the closure of A' (i.e., A'^*), which saves some computational effort. The projection itself can be computed using the A^* matrix of the underlying metagraph. Although the complexity of the procedure is exponential in the number of triples in the relevant portion of the A^* matrix, the size of this portion depends upon the projection set (the elements in the generating set that define the projection). Since in practice this set is not large (otherwise, the benefit of constructing the projection is lost), the procedure is still practical.

Consider again the conditional metagraph illustrated in Figure 7.19. If we project this metagraph over the set of variables $X' = \{ADV, ECON, NI, UCOST, VOL\} \subset X_v$, the result is the conditional metagraph illustrated in Figure 7.22. Two propositions, *cadv* and *mkt*, do not appear in the projection. The

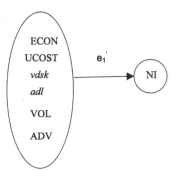

Figure 7.23. $N(\{ADV, ECON, NI, UCOST, VOL\}, K(\{inf, kv\}, \{cadv, mkt\}, S)) = K(\{inf, pkv\},$
$\{cadv, mkt\}, N(\{ADV, ECON, NI, UCOST, VOL\}, S)).$

reason is that *cadv* appears only in the invertex of e_2 and *mkt* appears only
in the invertex of e_6, and neither of these edges is a member of a dominant
metapath corresponding to the edge e_1'' in the projection.

We illustrate this commutativity property using our earlier example and the
context and projection on it, shown in Figures 7.21 and 7.22, respectively.
It should be easy to see that the projection of the conditional metagraph
in Figure 7.21 over $X' = \{ADV, ECON, NI, UCOST, VOL\}$ is the conditional
metagraph in Figure 7.23, which is also the result of defining the context
$P = \{inf, pkv\}, Q = \{cadv, mkt\}$ for the metagraph in Figure 7.22.

We have seen that metagraphs, as a tool for model management, can be
extended to incorporate assumptions about models. We have also defined a
context as a view of a model base, and have compared it to another type of
view, a projection. We now examine how these ideas can be applied to certain
questions that might arise about assumptions and their role in model management, and illustrate these applications using the example in Figure 7.19.

The first question is: given a model base represented by a conditional metagraph S, a source set of variables B whose values are known, a target set of
variables C whose values are to be computed, a set of propositions P that are
known to be true and a set of propositions Q that are known to be false, can we
determine the values of the variables C? This problem can be formulated as:
is there a metapath from B to C in the context $K(P, Q, S)$? Equivalently, we
can ask, if we project $K(P, Q, S)$ over $(B \cup C)$, is there an edge (B', C) (with
$B' \subseteq B$) in the projection? Since context and projection are commutative, this
is the same as asking whether, if we project S over $B \cup C$ and then construct
the context of the projection over (P, Q), the resulting metagraph has an edge
(B', C). The solution to this problem consists of two steps, (1) construction
of the context metagraph K (and its closure A^*), and (2) finding a metapath
$M(B, C)$ in K. For example, in Figure 7.21, this analysis can be used to find

that given $ECON, UCOST, ADV$, and VOL, it is possible to compute NI in the context K.

A second, related question is, given the same information, if there is no such metapath, how can we modify the problem (without modifying the model base) so that the target elements can be calculated? Three possible modifications that can be pursued are as follows:

1. Expand the source set B by determining values for some additional variables. An analysis along these lines can help determine whether it is worth the additional effort needed to evaluate those additional variables. For example, in Figure 7.21 a failed search for a metapath from $ADV, ECON$, and $UCOST$ to NI can show (through the identification of coinputs of candidate triples in A_k^*) that knowledge of VOL and satisfaction of $vdsk$ will yield a valid metapath.

2. Identify those variables in C that are not computable given B, and consider the possibility of removing them from the target set C (for instance, if they are only marginally useful or not critical for the decision problem at hand). Such variables can be identified as a by-product of the initial procedure for searching for a metapath from B to C using A^*. For example, in Figure 7.21, if $B = \{ECON, UCOST, VOL\}$ and $C = \{EXP, REV\}$, then no metapath exists; however, if we remove EXP from C, then there is a metapath.

3. Modify the context, by removing one or more assumptions from Q, so that certain models earlier invalid can now be considered. This can be achieved by incrementally removing assumptions and sets of assumptions from Q and repeating the metapath search procedure described earlier. For example, if $B = \{ADV, ECON, EQT, UCOST, VOL\}$, and $C = \{STK\}$, then there is no metapath $M(B, C)$ in the context of Figure 7.21. However, if mkt is removed from Q, then a metapath can indeed be found.

A third question is as follows: given a model base, source and target sets B, C respectively, what assumptions are necessary for computing C? This problem can be formulated as the identification of the set of assumptions that cannot be in Q for a conditional metapath $M(B, C)$ to exist in $K(\varnothing, Q, S)$. This is a useful question in practice, since it identifies those assumptions that are critical for the given decision problem. In order to solve this, once again the adjacency matrix of the conditional metagraph can be exploited. For example, in Figure 7.20, the propositions adl, inf, pkv and $vdsk$ are critical (i.e., cannot be in Q) for a metapath $M(\{ADV, ECON, UCOST, VOL\}, \{NI\})$ to exist.

A fourth question, related to the previous one, is: given S, B, C as before, and a specific proposition p, is p a necessary assumption for C to be calculated from B? This is a more specific problem than the identification of all necessary assumptions, and can be addressed using a simpler procedure. For example, in Figure 7.18, if $B = \{EQT, EXP, PRI, VOL\}$, and $C = \{STK\}$, then the procedure could be used to show that *vdsk* is required (i.e., *vdsk* cannot be in Q) while *adl* is not essential (i.e., *adl* can be in Q) for a metapath $M(B, C)$ to exist.

Thus, a number of important questions regarding assumptions and models can be formulated as questions of connectivity in conditional metagraphs, or in specific views of such metagraphs. Furthermore, these problems can be solved by visual inspection of the pictorial representation of the metagraph if the model base (or at least the relevant portion of it) is small, as well as through structured procedures on the algebraic representation of the metagraph in situations which are more complex. These procedures can also be used as a basis of sensitivity analysis, by analyzing the effect of changing the truth value of one or more propositions. This implies reapplying the procedures with modified contexts.

It should be noted that this is not the only type of sensitivity analysis used in modeling. Another involves changing values of one or more inputs (metapath source variables) to the decision problem, and studying the resulting impact on the model outcomes (metapath target variables). Unlike the assumption-based analysis described above, this requires actual execution of the models themselves, with different values for the input variables. During this type of analysis, an intermediate assumption that was true for one set of inputs may become false for another. Metagraphs are also useful for such analyses. For instance, for each intermediate assumption in a metapath, one can determine which input variables might change the truth value of that assumption. That is, in the A^* matrix, given an input variable, and an intermediate assumption $x \in X_v$, if there is a triple $t \in a^*_{xp}$ (and therefore, a simple path from x to p), such that the edges in that path are in the metapath, then p can be affected by changes to the value of x; otherwise, p is not affected by x. For example, in Figure 7.20, we can determine that the input variable *UCOST* does not affect the intermediate assumption *pkv*, and thus changes to *UCOST* will not affect the validity of the metapath $\{e_1, \ldots, e_5, e_8, e_9\}$ from $\{ADV, ECON, UCOST, VOL\}$ to $\{NI\}$. On the other hand, the input variable *ADV* can change the truth value of *pkv* through a change in *VOL*, and thus a change in the value of *ADV* may result in this metapath becoming invalid.

Chapter 8

METAGRAPHS IN DATA AND RULE MANAGEMENT

We now extend the results of Chapter 7 to encompass two additional information structures. The first is data bases, in which each edge represents a data relation with the key attributes as invertex and content attributes as outvertex. The second information structure is rule bases in which each edge represents a production rule with the antecedent (as a conjunction of propositions) as invertex and the consequent (also as a conjunction of propositions) as outvertex.

A simple example illustrating the use of metagraphs in representing models, data files, and rules is diagrammed in Figure 8.1. Figure 8.1(a) illustrates a model constrained by a rule. Edge e_1 is a model that calculates *Profit* in the outvertex using *Sales* and *COGS* (cost of goods sold) in the invertex. How-

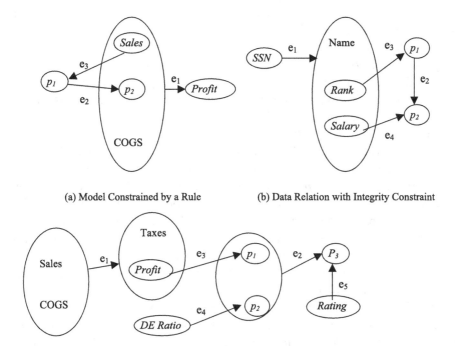

(a) Model Constrained by a Rule   (b) Data Relation with Integrity Constraint

(c) Rule Instantiated by a Model

Figure 8.1. Interactions between models, data, and rules.

ever, this model is valid only if proposition p_2 is valid. This proposition states simply that the model is valid. It is calculated from another proposition p_1, which might state, for example, that "*Sales* ≤ 1000", and the edge e_3 is used to calculate p_1 from the value of *Sales*. Edge e_2 is used to calculate p_2 from p_1 – that is, it states that if *Sales* ≤ 1000 then the model is valid.

We note that the rule represented by the edge e_2 does not tell us whether the model is invalid if *Sales* exceeds 1000. That is, it does not state that the model is valid if and only if *Sales* ≤ 1000 (i.e., $p_1 \equiv p_2$). At present, we are discussing only positive propositions. However, we will briefly discuss complementary propositions in Section 3 of this chapter.

Figure 8.1(b) illustrates a data relation with an integrity constraint. The edge e_1 represents the data relation with *SSN* (social security number) in the invertex as key attribute and *Name*, *Rank*, and *Salary* in the outvertex as content attributes. The constraint is that "IF *Rank* ≤ 5, THEN *Salary* $\leq \$20,000$". Edge e_2 represents this implication – that is, it states $p_1 \rightarrow p_2$, in which p_1 is "*Rank* ≤ 5" and p_2 is "*Salary* $\leq \$20,000$". The constraint states that if *Rank* ≤ 5, then *Salary* $\leq \$20,000$; it does not state what will happen if *Rank* > 5. Edge e_3 calculates proposition p_1 as a function of *Rank*, and edge e_4 calculates p_2 as a function of *Salary*. This example shows how the data values returned by a database query can be validated using any integrity constraints imposed upon them. The values of *Rank* and *Salary* corresponding to a given value of *SSN* in the database table represented by e_1 are valid only if they satisfy the constraint represented by the rule e_2.

The third example includes models, data, and rules. In this case, a model accesses a data relation to perform a calculation which is then used to instantiate a rule. This is illustrated in Figure 8.1(c). The edge e_1 represents a model that uses *Sales* and *COGS* (Cost of Goods Sold) to calculate *Taxes* and *Profit*. Edge e_3 is a rule that uses *Profit* to determine if proposition p_1 is true; p_1 is "*Profit* $\geq \$100,000$". Edge e_4 is a rule that uses *DE Ratio* (debt/equity ratio) to determine if proposition p_2 is true; p_2 is *DE Ratio* ≤ 1. Edge e_2 represents the rule $p_1 \wedge p_2 \rightarrow p_3$ – that is, "IF *Profit* $\geq \$100,000$ AND *DE Ratio* ≤ 1 THEN the company has an acceptably high rating". Edge e_5 presents an alternative method for determining if p_2 is true – that is, the acceptability of the rating can be determined directly from the rating itself. This example shows how rule e_2 can be used if the model e_1 is first used to determine a value for *Profit*, which is in turn used to evaluate p_1, one of the antecedents of the rule.

We can see that metagraphs may be used to model the three principal types of relationship found in decision support systems. These are data relations (in which case the elements in the generating set are data attributes), decision models (in which case the elements in the generating set are decision and other variables), and logical rules/constraints (in which case the elements in the generating set are Boolean variables or propositions). The purpose of this chapter is to investigate these topics in more detail. We will do so by examining three topics. The first is the representation of rule bases as metagraphs in which the elements in the metagraph correspond to propositions in the rule base. We will see that in an acyclic metagraph the existence of a metapath connecting two sets of elements is equivalent to the existence of an inference path connecting the corresponding propositions. The second topic is the use of metagraphs in integrating models, rules, and data. The third topic is the use of metagraphs in uncovering implicit integrity constraints in rule bases. This is done in Sections 1, 2 and 3 below.

1. REPRESENTING RULE BASES AS METAGRAPHS

We begin by examining the ways in which the connectivity properties (i.e., metapaths) and algebraic properties (i.e., the A^* matrix) of metagraphs can be useful in precompiling rule bases for efficient query processing. In this section, we will consider knowledge bases consisting only of rules, and in the following section we will consider the integration of rule bases with data and model bases.

When a metagraph is used to represent rule bases, each element in the generating set represents a proposition – that is, a variable that can take on either of two values, true or false. Each edge represents a rule in which the invertex is the antecedent to the rule and the outvertex is the consequent of the rule. The propositions in the antecedent are combined conjunctively as are the propositions in the consequent. Thus, a rule might be "IF the account balance is negative AND the amount is greater than \$1000 THEN send a dunning notice to the account AND notify the credit department". If $p_i, \ldots, p_4$ represent these propositions (e.g., p_1 is true *iff* the account balance is negative), then the rule would be written $p_1 \wedge p_2 \to p_3 \wedge p_4$. More generally, we define a rule base as follows:

DEFINITION 8.1. A *rule base* is an ordered pair $T = \langle P, R \rangle$ in which P is a set of propositions $\{p_i, i = 1, \ldots, I\}$ and R is a set of rules $R = \{r_k, k = 1, \ldots, K\}$ with each rule r_k being an expression of the form

$$\bigwedge_{p \in Y_k} p \to \bigwedge_{q \in Z_k} q,$$

with $Y_k \subseteq P$ and $Z_k \subseteq P$. Y_k is the antecedent of r_k and Z_k is the consequent.

We note that the rule base defined here can be used to represent only Horn clause rules, and that this limitation applies to all the results in this paper. Thus, we cannot represent rules of the form $\bigwedge_{p \in Y_k} p \to \bigvee_{q \in Z_k} q$. However,

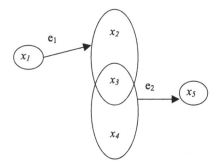

Figure 8.2. Metagraph representation of a rule base.

Horn clause logic has proven quite useful in rule based systems, and is not a serious practical limitation. The rule base defined above can be represented as a metagraph in which the following pairs are isomorphic: p_i and x_i, P and X, r_k and e_k, R and E, Y_k and V_k, and Z_k and W_k.

DEFINITION 8.2. Given a generating set X, a metagraph $S = \langle X, E \rangle$ on X with $E = \{e_k, k = 1, \ldots, K\}$ and a rule base $T = \langle P, R \rangle$ with $|X| = |P| = I$ and $|E| = |R| = K$ then S *corresponds to* T if for any $1 \leq i \leq I$ and $1 \leq k \leq K$, (1) $x_i \in V_k$ iff $p_i \in Y_k$, and (2) $x_i \in W_k$ iff $p_i \in Z_k$. If S corresponds to T, then X corresponds to P and E corresponds to R.

In the remainder of this chapter, we will use metagraph notation to refer to both metagraphs and rule bases. Thus, the metagraph of Figure 8.2 represents a rule base consisting of two rules: $x_1 \rightarrow x_2 \land x_3$ and $x_3 \land x_4 \rightarrow x_5$. We now establish a correspondence between metapaths and valid inferences.

THEOREM 8.1. *Let* $X = \{x_i, i = 1, \ldots, N\}$ *be a generating set and* $S = \langle X, E \rangle$ *be a metagraph on* X *corresponding to a rule base* $T = \langle P, R \rangle$. *Given two nonempty disjoint sets of elements* $X_1, X_2 \subset X$, *corresponding to two sets of propositions* $P_1, P_2 \subset P$, *there is an acyclic metapath* M *from* X_1 *to* X_2 *in* S *if and only if the following implication is valid in* T:

$$\bigwedge_{p \in P_1} p \rightarrow \bigwedge_{q \in P_2} q.$$

PROOF. (**IF**) Assume that there is a metapath M from B to C. Without loss of generality, assume that $|W_k| = 1, \forall_{e_k} \in M$. Now consider the following procedure:

Procedure Proof from $MP(X_1, X_2, M)$

Let $M_0 = M$, $X_0 = X_2$; $G = \emptyset$

While $X_0 = \emptyset$, DO
Step 1. Find $R \subseteq M_0$ such that $X_0 = \bigcup_{e_k \in R} W_k$.
Step 2. $G = G \cup \{(x, y) \mid (e_k \in R) \wedge (x \tilde{V}_k) \wedge (Y \in W_k)\}$.
Step 3. $X_0 = \bigcup_{e_k \in R} V_k \backslash \{X_1 \cup X_0\}$.
Step 4. $M_0 = M_0 \backslash R$.
Repeat;
End.

We prove the result by showing that the above procedure always terminates cessfully with a proof tree for X_2 from X_1. First, by definition of a metapath, we know that R can be found in step 1 in the first iteration. It follows then that X_2 can be inferred using R from the set of elements $\bigcup_{e_k \in R} V_k$. Step 2 identifies all leaf nodes in G that are not part of X_1, and then extends the proof tree backward from these elements. Since the metagraph is acyclic $W_k = 1$ for all $e_k \in M$, the rules considered for these must be distinct from the earlier set R. Also, since M is finite and in each iteration the candidate edge set is reduced, the procedure is guaranteed to terminate. The only possibility of unsuccessful termination is if $X_0 = 0$ in step 3 and $M_0 = 0$ in step 4 in the same iteration. However, this is impossible, since we know by definition that

$$\bigcup_{e_k \in M} V_k \backslash \bigcup_{e_k \in M} W_k \subseteq X_1$$

and thus, X_0 is made up of elements that have not been considered so far, and which, not being in X_1, must be outputs of rules in M that have not been considered (i.e., are in M_0). Thus, when the procedure terminates, it always yields a proof tree for X_2 in which all leaf nodes are elements of X_1, which is the desired result.

(ONLY IF) Assume that X_2 can be inferred from X_1. A set of elements X_2 can be inferred from another set of elements X_1 using a set of rules E if there is a proof tree whose non-leaf nodes are a superset of X_2, whose leaf nodes are a subset of X_1, and all of whose edges correspond to rules in E such that each non-leaf node x is the consequent of a rule whose antecedents are the children of x in the proof tree.

Since the metagraph is acyclic, every directed path to a non-leaf node can be extended to a leaf node that is part of X_1. Since every edge in the proof tree is based on an edge in the metagraph, if we consider the metagraph edges corresponding to all the edges in the proof, we get a set M such that every edge is on a path from X_1 to X_2. Furthermore, since all the leaf nodes are within X_1 and all elements of X_2 are non-leaf nodes, $X_2 \subseteq \bigcup_{e_k \in M} W_k$ and $\bigcup_{e_k \in M} V_k \backslash \bigcup_{e_k \in M} W_k \subseteq X_1$, so that M is indeed a metapath from X_1 to X_2.

Thus, in Figure 8.2 we can infer that $x_1 \wedge x_4 \rightarrow x_5$ because of the existence of a metapath $M(\{x_1, x_4\}, \{x_5\})$. Similarly, in the metagraph of Figure 8.3

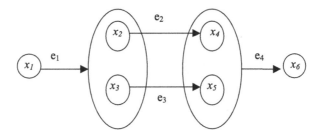

Figure 8.3. Metagraph with acyclic metapath.

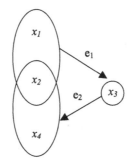

Figure 8.4. Cyclic metagraph.

we can infer $x_1 \rightarrow x_6$ because of the existence of a metapath $M(\{x_1\}, \{x_6\})$. On the other hand, in the cyclic metagraph of Figure 8.4 there is a metapath $M(\{x_1\}, \{x_2, x_3, x_4\})$, but we cannot infer $x_1 \rightarrow x_2 \wedge x_3 \wedge x_4$. For example, if we have x_1 true and x_2, x_3, x_4 false, then the rules corresponding to e_1 (i.e., $x_1 \wedge x_2 \rightarrow x_3$) and e_2 (i.e., $x_3 \rightarrow x_2 \wedge x_4$) are both true, but $x_1 \rightarrow x_2 \wedge x_3 \wedge x_4$ is false.   □

We now demonstrate that the construction of A^* can be useful in determining whether a set of propositions (denoted $B \subseteq X$) is sufficient to infer a second set of propositions (denoted $C \subseteq X$). This can be accomplished by means of a localized search involving only those triples in A^* contained in members a_{ij}^* for which $x_i \in B$ and $x_j \in C$.

THEOREM 8.2. *Let X be a generating set and $S = \langle X, E \rangle$ be a metagraph on X. Given two nonempty disjoint sets of elements $B, C \subseteq X$ and the set of all simple paths $\theta = \{h_1, h_2, \ldots, h_q\}$ from any element $x \in B$ to some $y \in C$, if M is a metapath from B to y, then $\exists H \in 2^\theta$, where 2^θ is the power set of θ, such that $M = Set(H)$.*

PROOF. By definition, every edge e_i in a metapath $M(B, C)$ must lie on a simple path h_i from some element $x \in B$ to some $y \in C$, where $Set(h_i) \subseteq$

$M(B, C)$. That is, M is made up of edges that comprise a set of simple paths $H_i = \{h'_1, \ldots, h'_t\}$ from elements in B to elements in C (i.e., $M = \bigcup_{i=1}^{t} Set(h'_i)$). Clearly, since θ is the set of all such paths, $H_1 \subseteq \theta$, and the results follow. □

Thus, the search for a metapath from B to C can be limited to unions of the edges in simple paths from elements in B to elements in C. All such paths are contained in the triples contained in the members a^*_{ij} for which $x_i \in B$ and $x_j \in C$. The following theorem simplifies this task even further.

THEOREM 8.3. *Given q paths $h_1, \ldots, h_q$ from a set of elements B to some set of elements C, let $\alpha_i = \bigcup$ (invertices on ith path) and $\beta_i = \bigcup$ (outvertices on ith path). If*

$$\bigcup_{i=1}^{q} \alpha_i \setminus \bigcup_{i=1}^{q} \beta_1 \subseteq B,$$

then $\bigcup_{i=1}^{q} Set(h_i)$ is a metapath from B to C.

PROOF. Follows from the definition of a metapath. □

For example, consider the metagraph of Figure 8.3 and it's adjacency matrix in Figure 8.5. From the closure A^* in Figure 8.6 we can determine whether $x_1 \rightarrow x_6$ by examining the triples in α^*_{16}. There are two such triples, corresponding to $\langle e_1, e_2, e_4 \rangle$ and $\langle e_1, e_3, e_4 \rangle$, with (α, β) components as follows:

First triple: $\alpha_1 = \{x_3\}, \beta_1 = \{x_2, x_3, x_4\}$.
Second triple: $\alpha_2 = \{x_4\}, \beta_2 = \{x_2, x_3, x_5\}$.

Since $(\alpha_1 \cup \beta_2) \setminus (\beta_1 \cup \beta_2) = \emptyset \in \{x_1\}$, we have $x_1 \rightarrow x_6$.

In this section, we have assumed that the metagraphs under consideration are acyclic. We note that a metapath constructed from acyclic simple paths need not be acyclic. Consider the metagraph of Figure 8.7, and let $M(\{x_1\}, \{x_4, x_5\}) = Set(h_l) \cup Set(h_2)$, where $h_l(x_1, x_5) = \langle e_1, e_2, e_3 \rangle$ and $h_2(x_1, x_4) = \langle e_4, e_5, e_6 \rangle$. Both of these paths are acyclic; but the metapath, which consists of all six edges, is cyclic. Even so, we can prove that $x_1 \rightarrow x_4 \wedge x_5$, since there is an acyclic metapath $\{e_1, e_3, e_4, e_6\}$ from $\{x_1\}$ to $\{x_4, x_5\}$.

These results may be very helpful in the processing of large rule bases for which there are likely to be a variety of queries, since A^* provides a precompilation of the rule base. Of course, the computational effort needed to answer any query will depend on the number of triples in the intersection members of A^*, but in rule bases where the number of paths between any two elements is

	x_1	x_2	x_3	x_4	x_5	x_6
x_1	∅	$\{\langle\emptyset, \{x_3\}, \langle e_1\rangle\rangle\}$	$\{\langle\emptyset, \{x_2\}, \langle e_1\rangle\rangle\}$	∅	∅	∅
x_2	∅	∅	∅	$\{\langle\emptyset,\emptyset, \langle e_1\rangle\rangle\}$	∅	∅
x_3	∅	∅	∅	∅	$\{\langle\emptyset, \emptyset, \langle e_3\rangle\rangle\}$	∅
x_4	∅	∅	∅	∅	∅	$\{\langle \{x_5\}, \emptyset, \langle e_4\rangle\rangle\}$
x_5	∅	∅	∅	∅	∅	$\{\langle \{x_4\}, \emptyset, \langle e_4\rangle\rangle\}$
x_6	∅	∅	∅	∅	∅	∅

Figure 8.5. The adjacency matrix of the metagraph in Figure 8.3.

	x_1	x_2	x_3	x_4	x_5	x_6
1	∅	$\{\langle\emptyset, \{x_3\}, \langle e_1\rangle\rangle\}$	$\{\langle\emptyset, \{x_2\}, \langle e_1\rangle\rangle\}$	$\{\langle\emptyset, \{x_3\}, \langle e_1, e_2\rangle\rangle\}$	$\{\langle\emptyset,\{x_2\},\langle e_1, e_3\rangle\rangle\}$	$\{\langle\{x_5\}, \{x_2, x_4\}, \langle e_1, e_2, {}_4\rangle\rangle,$ $\langle\{x_4\}, \{x_3, x_5\}, \langle e_1, e_3, {}_4\rangle\rangle\}$
x_2	∅	∅	∅	$\{\langle\emptyset, \emptyset, \langle e_1\rangle\rangle\}$	∅	$\{\langle\{x_5\}, \{x_4\}, \langle e_2, e_4\rangle\rangle\}$
x_3	∅	∅	∅	∅	$\{\langle\emptyset, \emptyset, \langle e_3\rangle\rangle\}$	$\{\langle\{x_4\}, \{x_5\}, \langle e_3, e_4\rangle\rangle\}$
x_4	∅	∅	∅	∅	∅	$\{\langle\{x_5\}, \emptyset, \langle e_4\rangle\rangle\}$
x_5	∅	∅	∅	∅	∅	$\{\langle\{x_4\}, \emptyset, \langle e_4\rangle\rangle\}$
x_6	∅	∅	∅	∅	∅	∅

Figure 8.6. The closure of the adjacency matrix in Figure 8.5.

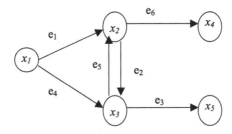

Figure 8.7. Metagraph with cyclic and acyclic metapaths.

not very large, precompilation of A^* may facilitate localized search for solutions to inference problems.

2. INTEGRATING RULES, MODELS AND DATA

We have already discussed the need to support different resources such as models, rules and data in a DSS. In this section, we discuss how metagraph representation of these resources can be useful both during DSS design and use. To facilitate our discussion, we use the term "knowledge base" to describe the structure containing all these resources in the DSS. In other words, the knowledge base is the combination of the model base, rule base and data base contained in a DSS.

We start by defining a metagraph for a DSS knowledge base. The generating set X for such a metagraph consists of two types of elements. That is, $X = \{X_p \cup X_v\}$, where X_p is a set of propositions, and X_v is a set of variables. The basic difference between a variable and a proposition is that the latter is a logical variable restricted to a truth value (in this paper, we will assume a Boolean truth value), while the value space for a variable can be any arbitrary set.

The edges of the metagraph then correspond to the different resource modules, namely models, rules and data. In the case of a model edge, the invertex consists of the inputs and assumptions, while the outvertex identifies the output variables. A data edge represents a functional dependency between the invertex elements (key) and the outvertex elements (non-key/content). A number of significant classes of edges can be characterized, based on the above partitioning of the generating set:

1. If $V(e) \subseteq X_p$ and $W(e) \subseteq X_p$, then e is a *rule*.
2. If $V(e) \subseteq X_v$ and $W(e) \subseteq X_p$, then e is a *proposition definition*.
3. If $V(e) \subseteq X_v$ and $W(e) \subseteq X_v$, then e is an *unconstrained model* or *data relation*.
4. If $V(e) \subseteq X_p$ and $W(e) \subseteq X_p$, then e is an *equality predicate*.
5. If $V(e) \subseteq X$ and $W(e) \subseteq X_v$, then e is a *constrained model* or *data relation with domain constraints*.

In order to understand the significance of the above characterizations, we present the following observations:

- The classification is not meant to be exhaustive. For instance, it is quite possible for an edge representing a model to have both variables and propositions in either or both vertices. The primary objective of the classification is to identify certain classes that may be significant in metagraph-based analysis during system design and use.
- Even though it is possible to design modules that include features of several of the above classes, it is a good idea to design modules such that each module belongs to one of these classes. The reason for this is that it facilitates easier implementation of modules. For instance, if rules are

constrained to propositions only, then they can be implemented by a simpler knowledge representation construct, and used with a simpler inference engine. If, during requirements determination, modules are identified that combine features of multiple classes, it may often be desirable to decompose such modules into sets of simpler modules that do conform to the classes. The functionality of the original complex module can then be recreated through appropriate module integration during problem solving.

- Even though the different types of modules are represented using a common construct, a metagraph edge, their semantics are quite different. For instance, rule edges are logical implications; thus, the truth value of the outvertex propositions (consequent) is determined only when the invertex propositions (antecedent) are all true. In general, whenever one or more propositions occur in the invertex of an edge, the relationship underlying the edge is interpreted as being conditional upon these propositions being true. For instance, in the case of a constrained model edge, the underlying model can be used to compute the outvertex variables from the invertex variables, as long as the propositions in the invertex are true. Similarly, for a range restricted data relation, the corresponding edge represents a functional dependency, which can be used as an integrity constraint upon the database. That is, the functional dependency can be used to validate updates, and also to construct data access plans in query processing. Note that the functional dependency can be used to access data only when the specified invertex variables are assigned values that correspond to existing attribute values in the current extensional database.

- An equality predicate is an edge that can be used to assign a value to a variable, given the value of one or more propositions. This type of edge is likely only when the invertex consists of a single proposition that is an assignment statement and the outvertex is a single variable. For instance, the proposition p_i "the company is of type X" can be used to evaluate the value of the variable *company type*, using the edge $\langle \{p_i\}, \{company\ type\} \rangle$. However, the proposition "$A \geq 5$" is insufficient to compute the value of A (of course, it could be combined with another proposition "$A \leq 5$" to compute A, but such edges are not likely to occur very often).

We now discuss an application of the metagraph representation. We show how a metapath can be used as a basis for identifying collections of resource modules that can be used to solve certain problems.

Consider a situation where a user wants to obtain values for a set C of elements (variables and/or propositions), given values for another set B of elements. Depending on the available knowledge base, there may be a number of possible solution plans that can achieve this. When the knowledge base consists of models and data, potentially feasible solution plans can be identified by finding metapaths from the set B to the set C in the metagraph representation. Also, for a rule base, the search for inference plans for deducing C from B can also be formulated as a search for an acyclic metapath from B to C. We now examine the general case when the knowledge base contains all three types of components.

While the general intuition still holds in this case, that a metapath represents a potentially feasible solution plan, the issue of cycles has to be considered. Recall that in a rule base, a metapath is usable as a basis for inference only when it is acyclic. On the other hand, in a model and/or database, there are no such constraints. For instance, the existence of a cycle in a metapath corresponding to an integrated model (a collection of models for a given B, C pair), merely denotes a set of variables that have to be equilibrated through possible repetitive iteration through the edges in the cycle. In other words, the existence of a cycle in a metapath does not invalidate it in the case of an integrated model, but does present problems in the case of a rule-based inference process. What then is the situation when the metapath corresponds to a combination of rule-based inference, model execution and data access?

In order to address this question, we need to define some additional terms:

DEFINITION 8.3. Given a metapath $M(B, C)$ in a metagraph, and an element $x \in X$, we say that x is *cyclic within* M if there is a cycle $h(x, x)$ such that $Set(h(x, x)) \subseteq M$.

DEFINITION 8.4. An element x is said to be *used* in a metapath $M(B, C)$ if $\exists e \in M$ such that $x \in \{V(e) \cup W(e)\}$.

We now have the following result which extends our earlier work, for the general case:

THEOREM 8.4. *Given a metapath $M(B, C)$ consisting of model, rule and data edges, these modules can be used to compute values for the elements of C, given specific values for the elements of B, if none of the propositions used in M are cyclic within M.*

PROOF. It has been shown in Chapter 2 that the computability of a set of variables C from a given set of inputs B using a set of data relations and models, can be determined by the existence of a metapath from B to C. Thus, we only need to examine the case where the available knowledge base includes rules. However, by Theorem 8.1, we know that for a rule base, a set of propositions C can be inferred from another set B if there is an acyclic metapath from B

to C (i.e., if all the propositions in the metapath are acyclic within it). Thus, the result follows.   □

We can see from Chapter 2 that a metapath consisting of model and data edges is sufficient to compute C given B. This follows from the definition of a metapath, and is independent of whether the metapath is cyclic or acyclic. Theorem 8.4 shows that rule edges can be used in the metapath as well, provided that there are no cycles through the propositions in these rules.

Although the qualification about cyclic propositions in Theorem 8.4 is significant, it can be handled quite conveniently in practice. This is because of the A^* matrix, which enables the necessary check to be performed very easily, in two steps. First, for each proposition used in the metapath, we can determine whether it is cyclical by examining the diagonal cell of A^* corresponding to that proposition. Clearly, if that cell is empty, then the proposition is acyclic. Second, if the diagonal cell is non-empty, we can check whether any of the triples in that cell corresponds to a path that is contained entirely within M. If not, then the proposition is still not cyclic within M, and the latter can be used to construct a solution plan.

We now present an example, to illustrate the use of metagraph representations of DSS knowledge bases. Consider a DSS knowledge base consisting of the following modules:

e_1: **Rev, Trate, Exp** → **NI** (this is an accounting model that computes net income, given revenue, expenses and the applicable tax rate).

e_2: **Pri, Econ** → **Vol** (this is a marketing model that determines volume demanded, given unit price and the value of an economic indicator).

e_3: **Vol** → **Exp** (this is a simple accounting model that computes expenses as a function of volume of sales).

e_4: **Pri, Vol, Dis** → **Rev** (this is a marketing model that computes revenue for any given level of price, volume and discount rate).

e_5: p_1, p_2 → p_3 (this is a rule stipulating that the applicable discount rate is 10% if the item's unit price is less than or equal to \$100 and the volume demanded is greater than or equal to 1000 units – i.e., $p_1 \wedge p_2 \to p_3$).

e_6: p_3 → **Dis** (this is an equality predicate that allows the assignment of a value, 10%, to the discount rate variable when the corresponding proposition is true. Note that the converse relationship also holds, that is, given a value for the discount rate, the proposition "Dis $= 10\%$" can be evaluated as well).

e_7: **Vol, Exp** → **Pri** (this is a pricing model used to determine the optimal price for any given level of volume and expenses).

e_8: **Pri** → p_1 (this module defines the proposition "Pri $\leq$ \$100" from the variable **Pri**).

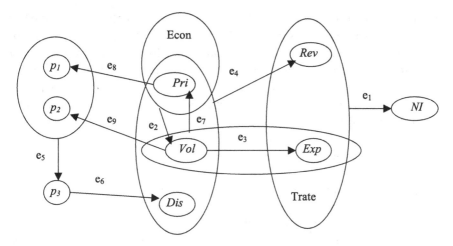

Figure 8.8. Metagraph representation of a DSS knowledge base.

e_9: **Vol** $\rightarrow$ p_2 (definition of the proposition "Vol $\geq$ 1000" from the variable **Vol**).

The interpretation of the propositions is as shown in Figure 8.8, which shows a metagraph representation of the knowledge base. Analysis of this metagraph and its adjacency matrix A (and the corresponding closure A^*) leads to the following conclusions:

- If price is known, and if net income is the variable that we want to evaluate, there are a number of paths from price to net income (e.g., $\langle e_4, e_1 \rangle$, $\langle e_2, e_3, e_1 \rangle$, $\langle e_8, e_5, e_6, e_4, e_1 \rangle$). In each case, there are additional coinputs for which values have to be determined. For instance, in the path $\langle e_4, e_1 \rangle$, the coinputs of price are volume, discount rate and expenses.
- There is a metapath $M_1 = \{e_1, e_2, e_3, e_4, e_5, e_6, e_8, e_9\}$ from price and economic indicator to net income. In other words, given values of the first two variables, we can compute the corresponding value of net income. However, this is only true under certain conditions. In particular, this is only true when the rule e_5 fires successfully. Thus, in this case, the rule in effect establishes the range of application of the metapath as a viable solution basis.
- The metapath M_1 is viable, even though there is a cycle within it, consisting of the path $\langle e_2, e_8 \rangle$. However, this cycle does not invalidate the metapath, since the cyclical elements in this case (price and volume) are both variables, and not propositions. Thus, the cycle merely indicates that the values of price and volume have to be equilibrated (perhaps through repeated iteration through the cycle) as part of the solution process.

- The set of propositions used in the metapath, whether in rules or as constraints in models, serve to define the range of application of the integrated model or solution plan represented by the metapath. In other words, we can infer that the metapath is usable only when prices is below $100, volume is greater than 1000 units, and the discount rate is 10%. Note that only those propositions that occur in the invertices of edges have to be considered. Thus, for instance, if the outvertex of e_5 included a proposition "sales terms are net 50", this proposition would not be a requirement for the metapath M_l to be usable.
- If the discount rate is also known (and it need not be 10%), then there is also the metapath $M_2 = \{e_l, e_2, e_3, e_4, e_7\}$ from price, economic indicator and discount rate to net income. Unlike M_1, this metapath represents an unconstrained model. In other words, it can be used for any values of the input variables. However, the rule e_5 can still serve a valuable purpose when M_2 is used, namely as an integrity constraint. Thus, it can be used to check whether the values of price, volume and discount rate are consistent, in cases where the rule is applicable. Note that in order to utilize e_5 in this way, we have to use e_6 in the reverse direction, which is acceptable for an equality predicate, as mentioned earlier.

The above observations imply that metagraph representation of a DSS knowledge base facilitates discovery of a variety of useful information. An important point that must be stressed here is that although many of these observations can be made by visual analysis, which might be practical for small knowledge bases, the strength of the metagraph approach lies in the fact that these conclusions can also be reached through analysis of the algebraic representation of the metagraph through its A and A^* matrices. This is important, since it enables a metagraph-based DSS to not only help a decision maker by providing an expressive graphical visualization tool, but also provides active support through analytical processes that are transparent to the user but still provide valuable evaluative information about the available resources and potential solution plans.

The use of the adjacency and closure matrices to find metapaths between specific sets of elements has been discussed in the previous section. We next show how these matrices can also be exploited to find rules that may be applicable as integrity constraints on metapaths representing problem solutions.

To start with, note that any rule such that all its elements are reachable from a set of elements is a potential integrity constraint upon that set of elements. For example, in Figure 8.8, the rule corresponding to e_5 is a potential integrity constraint upon any metapath containing e_4 since the elements p_1, p_2, p_3 in e_5 are all reachable from the elements of e_4. The simplest case is when the elements of a rule r are defined in terms of the elements in the metapath M. This case can be easily identified as follows:

1. For each variable $x \in X_v$ occurring in M, find all the propositions $y \in X_p$ that are reachable from x in A (i.e., such that $a_{xy} \neq \emptyset$). Let this set of propositions be X'_p.
2. For each proposition $x \in X'_p$, examine a_{yz} for any triple t such that $z \in X'_p$ and the coinput of y in t is contained in X'_p. If any such triple is found, then $edge(t)$ is a potential integrity constraint.

A more general case is where the potential integrity constraint is indirectly reachable from the elements in M via a metapath consisting of rules. This case can be checked using the metapath search procedure itself, restricting the search to metapaths consisting solely of rule edges (i.e., triples for which all coinputs and outputs are propositions).

3. DISCOVERING IMPLICIT INTEGRITY CONSTRAINTS

In the previous sections, we have assumed that all the rules in the rule base represented by a metagraph contain only positive propositions. In this section, we discuss how information about complementary literals can be used to make certain useful transformations in the metagraph representation of a rule base that can reveal relevant integrity constraints.

The issue of integrity maintenance is an important one in the context of knowledge based systems and information systems in general. One way in which integrity is enforced in a rule based Knowledge Based Systems (KBSs) is through the use of integrity constraints. While any sentence can be used as an integrity constraint, a common and fairly widely applicable form for integrity constraints is a statement of the form:

$$\neg x_1 \vee \neg x_2 \cdots \vee \neg x_n$$

which can also be stated in the form

$$x_1 \wedge x_2 \cdots \wedge x_{k-1} \wedge x_{k+1} \cdots \wedge x_N \to \neg x_k$$

for any k between 1 and N. Typically, as part of the definition of the rule base, a number of such integrity constraints may be included. These constraints can be used not only during the problem solving process to eliminate infeasible solutions, but also as integrity checks during any updates to the rule base (Grant and Minker, 1991).

In addition to the explicitly stated integrity constraints in the rule base, implicit constraints may exist in the knowledge base, or may result from determinations that some propositions are either complementary or mutually contradictory.

EXAMPLE. Consider the following rule base: $\{(\alpha \to p), (\alpha \to q)\}$, where α is some conjunction of propositions containing neither p nor q.

In this rule base, if we impose an explicit integrity constraint $(\neg p \vee \neg q)$, we find that there is an implicit integrity constraint $\neg \alpha$ which the system designer or user may not be aware of.

Discovery of such implicit integrity constraints can be very useful during problem solving. Methods for the discovery of such implicit constraints are valuable, since they facilitate rule base management. In this section, we show that metagraph representation of rules and the corresponding A^* matrix can facilitate the discovery of implicit integrity constraints. For this purpose, we need to define an augmented form of the triples in A^*. Given a triple $\langle \{\alpha\}, \{\beta\}, \langle h \rangle \rangle$ in a_{ij}^* (where α and β are sets of elements and h is a sequence of edges), the augmented triple corresponding to this is given by $\langle \{\alpha, x_i\}, \{\beta, x_j\}, \langle h \rangle \rangle$.

THEOREM 8.5. *Given a rule base and its corresponding metagraph S, if the constraint $\neg x_i \vee \neg x_j$ is added, then the following rules of transformation can be used to simplify the A^* matrix of S (where $\alpha, \beta \subset X$), and $x_i, x_j \in X$ and $\{x_i, x_j\} \cap \{\alpha \cup \beta\} = \phi$ and where h is a sequence of edges forming a simple path):*

1. *Any augmented triple of the form $\langle \{x_i, x_j, \alpha\}, \{\beta\}, \langle h \rangle \rangle$ can be deleted;*
2. *From any augmented triple of the form $\langle \{x_i, \alpha\}, \{x_j, \beta\}, \langle h \rangle \rangle$ the integrity constraint $\neg \alpha \vee \neg x_i$ can be inferred;*
3. *From any augmented triple of the form $\langle \{\alpha\}, \{x_i, x_j, \beta\}, \langle h \rangle \rangle$ the integrity constraint $\neg \alpha$ can be inferred.*

PROOF. We consider each rule in turn:

Transformation 1: Given the integrity constraint, clearly both x_i and x_j can never be both true, so that the antecedent of the rule is never true, and thus the rule is useless and can be deleted.

Transformation 2: Triples of this type correspond to the implication $\alpha \wedge x_i \to \beta \wedge x_j$ which can be simplified to the two clauses

$$\alpha \wedge x_i \to \beta,$$

$$\alpha \wedge x_i \to x_j.$$

The second clause can be resolved with the integrity constraint to infer the integrity constraint $\neg \alpha \vee \neg x_i$, which is the desired result.

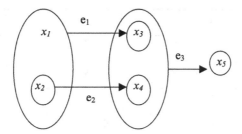

Figure 8.9. Metagraph illustrating integrity constraints.

Transformation 3: Triples of this type correspond to the implication $\alpha \rightarrow$ $\beta \wedge x_i \wedge x_j$. However, the given integrity constraint is equivalent to the expression $\neg(x_i \wedge x_j)$. Hence the consequent of the rule must always be false. This in turn implies that a must always be false (i.e., $\neg\alpha$), which is the desired result. □

The transformations in Theorem 8.5 can be used to eliminate some triples (Transformation 1) and to extract implicit integrity constraints (Transformations 2 and 3). We illustrate this in the following example:

EXAMPLE. Consider the metagraph in Figure 8.9, it's a matrix in Figure 8.10, and its A^* matrix in Figure 8.11. The following separate examples illustrate each of the three rules in Theorem 8.5:

If we introduce the constraint $\neg x_1 \vee \neg x_4$, then the triple $\langle\{x_1, x_4), \{x_3\},$ $\langle e_1, e_3\rangle\rangle$ in a_{25}^* can be deleted by Rule 1.
If we introduce the constraint, then the in $\neg x_2 \vee \neg x_3$, then the integrity constraint $\neg x_1 \vee \neg x_2 \vee \neg x_4$ can be inferred by Rule 2.
If we introduce the constraint $\neg x_4 \vee \neg x_5$, this leads to the constraint $\neg x_2 \vee$ $\neg x_3$ by Rule 3.

Furthermore, if we consider not only the simple paths represented by the triples in A^*, but also metapaths, then we get the following additional transformation:

THEOREM 8.6. *Given a rule base, if the constraint $\neg x_i \vee \neg x_j$ is added, then for any set of elements α such that there is an acyclic metapath in the corresponding metagraph from α to both x_i and x_j, the integrity constraint $\neg\alpha$ can be inferred.*

PROOF. From the previous section, we know that if there is a metapath from α to both x_i and x_j, then it follows that

$$\alpha \rightarrow x_i \wedge x_j.$$

	x_1	x_2	x_3	x_4	x_5
x_1	$\varnothing$	$\varnothing$	$\{\langle\{x_2\}, \varnothing, \langle e_1\rangle\rangle\}$	$\varnothing$	$\varnothing$
x_2	$\varnothing$	$\varnothing$	$\{\langle\{x_1\}, \varnothing, \langle e_1\rangle\rangle\}$	$\{\langle\varnothing, \varnothing, \langle e_2\rangle\rangle\}$	$\varnothing$
x_3	$\varnothing$	$\varnothing$	$\varnothing$	$\varnothing$	$\{\langle\{x_4\}, \varnothing, \langle e_3\rangle\rangle\}$
x_4	$\varnothing$	$\varnothing$	$\varnothing$	$\varnothing$	$\{\langle\{x_3\}, \varnothing, \langle e_3\rangle\rangle\}$
x_5	$\varnothing$	$\varnothing$	$\varnothing$	$\varnothing$	$\varnothing$

Figure 8.10. The A matrix of the integrity constraint metagraph.

	x_1	x_2	x_3	x_4	x_5
x_1	$\varnothing$	$\varnothing$	$\{\langle\{x_2\}, \varnothing, \langle e_1\rangle\rangle\}$	$\varnothing$	$\{\langle\{x_2, x_4\}, \{x_3\}, \langle e_1, e_3\rangle\rangle\}$
x_2	$\varnothing$	$\varnothing$	$\{\langle\{x_1\}, \varnothing, \langle e_1\rangle\rangle\}$	$\{\langle\varnothing, \varnothing, \langle e_2\rangle\rangle\}$	$\{\langle\{x_1, x_4\}, \{x_3\}, \langle e_1, e_3\rangle\rangle,$ $\langle\{x_3\}, \{x_4\}, \langle e_2, e_3\rangle\rangle\}$
x_3	$\varnothing$	$\varnothing$	$\varnothing$	$\varnothing$	$\{\langle\{x_4\}, \varnothing, \langle e_3\rangle\rangle\}$
x_4	$\varnothing$	$\varnothing$	$\varnothing$	$\varnothing$	$\{\langle\{x_3\}, \varnothing, \langle e_3\rangle\rangle\}$
x_5	$\varnothing$	$\varnothing$	$\varnothing$	$\varnothing$	$\varnothing$

Figure 8.11. The A^* matrix of the integrity constraint metagraph.

However, this corresponds to the case where Transformation 3 applies in Theorem 8.5, and the result follows.   □

EXAMPLE. Consider yet again the metagraph in Figure 8.9, and its corresponding A and A^* matrices (Figures 8.10 and 8.11). The metapath $\{e_1, e_2\}$ connects the set $\{x_1, x_2\}$ to the set $\{x_3, x_4\}$. If we add the constraint $\neg x_3 \vee \neg x_4$, then by Theorem 8.6, we get the integrity constraint $\neg x_1 \vee \neg x_2$ (it is worth noting that in this case, we also find that by Transformation 1, the edge e_3 can be deleted).

Thus, we have shown that the A^* matrix for a metagraph corresponding to a rule base provides a useful basis for not only determining valid, but also for detecting implicit integrity constraints. Furthermore, since the set of integrity constraints only changes when the rule base is updated, the process of extracting the constraints can be viewed as part of the compilation of the rule base in terms of the A^* matrix.

4. METAGRAPH MODELS OF DECISION SUPPORT SYSTEMS

In Chapter 7 we found that metagraphs provide a useful way of modeling decision models. The elements represented the input and output variables in the model, and the edges represented the models themselves, with the model inputs as invertices and the model outputs as outvertices. The collection of set-to-set mappings that make up a metagraph provide a useful framework for model management.

In this chapter, we have seen that the same is true of data relations and rules. In the case of data relations, the elements in the generating set are data attributes, and the edges are the relations, with the invertices as key attributes and the outvertices as content attributes. In the case of rules, the elements in the generating set are propositions and the edges are the rules, with the invertices as antecedents and the outvertices as consequents, both in conjunctive form. In both cases the collection of set-to-set mappings that make up a metagraph provide a useful framework for the management of these two additional types of information found in a decision support system.

We can also see that the concept of a metapath provide a useful framework for analysis in these additional areas as well. Metapaths can be used not only to identify calculation paths for models, but also to identify access paths for data retrieval and inference paths for collections of rules. As with model bases, there must be an acyclic metapath for the inferences to be valid. In addition, since the same structure (metagraphs and acyclic metapaths) can be used to determine inferences in each of these three areas, they can be useful in collections of data, models, and rules.

The key to this is in the algebraic foundation of metagraphs, specifically the adjacency matrix and its closure. Their principal contribution is in identifying inference paths and in determining whether any such paths are acyclic. In addition, when the elements in the generating set are propositions, they can be used to help analyze integrity constraints, which include the case the propositions may be negated. Thus, metagraphs and their algebraic underpinnings provide a powerful framework for the representation and algebraic analysis of the three principal types of information found in a DSS – models, data relations, and rules.

Chapter 9

METAGRAPHS IN WORKFLOW AND PROCESS ANALYSIS

This is the last of the three chapters in which we examine the applications of metagraphs to information processing systems. In the previous two chapters we examined applications to three information structures found in decision support systems: data, models, and rules. We now turn to yet another topic – workflow systems. Workflow systems integrate the judgmental and decision making efforts of humans (managers and analysts) with the information processing (computational and communication) activities of machines to implement business processes.

Business processes – such as order fulfillment, product development, corporate budgeting, and interorganizational supply chain management – cut across traditional organizational functions – such as purchasing, manufacturing, distribution, and marketing. Organizational structures are often best described by hierarchical or matrix decomposition of functions. People are typically recruited, trained, and advanced within a specialized function, at least until they are advanced near or to the pinnacle of the organization. On the other hand, organizational goals often are accomplished by cross-functional business processes, especially those processes that interface the organization with its customers. Workflow systems crosswalk functions with processes so that organizational structures may be employed to accomplish organizational goals.

Metagraphs provide a useful tool for modeling workflows and their underlying processes. The elements in the generating set are the objects being processed by the workflow. Often these are documents, such as loan applications, credit reports, property data, and risk analysis reports. We will not focus here on the detailed content of these reports but rather on their flow among tasks, each of which transforms one set of documents into another set of documents. These tasks may include risk assessment, property appraisal, and approval/rejection of a proposal. The tasks are represented by edges in a metagraph.

In this chapter we will address five issues relevant to the application of metagraphs to workflow and process analysis. The first is the use of metagraphs in representing workflows and processes. A process is represented by a conditional metagraph in which the propositions have not been evaluated (i.e., assigned a TRUE or FALSE designation). For example, a proposition associated with an edge may be an assumption that the task represented by the edge exceeds a minimum dollar amount. In addition resources (e.g., specialized people or equipment) may be associated with tasks, and a proposition may be that sufficient resources are available. A workflow is an instantiation of the processes – that is, a process in which the propositions have been evaluated.

The second issue is the ways in which views of a metagraph – that is, projections of workflows – may be used to identify information interactions in the workflow. These include task interactions, in which one task produces an information element (in its outvertex) that is used in another task (in its invertex), and resource interactions. Both task and resource interactions can be represented by metagraphs as well. We will introduce two new metagraph views, the Task Interaction Metagraph and the Resource Interaction Metagraph, to represent these two views of a workflow.

The third issue concerns the synthesis of processes, which is accomplished by taking the union of the metagraphs representing the processes. An interesting question is whether full connectivity is preserved. If it is not, then there is at least one instantiation (i.e., a workflow) of the synthesized process that cannot be completed. We are also concerned with redundancy and full connectivity in the synthesized process. We will see that an important issue here is whether the process contains one or more cycles.

The fourth issue concerns process decomposition and its implications for organizational design. In this case an important consideration is independence of decomposed subprocesses and the resources used in these subprocesses (as specified in the Resource Independence Metagraph). The organizational design issues result from the dependence or independence of the submetagraph representing the subprocess and the resources used in the subprocess.

The fifth issue differs from the first four in that it concerns quantitative rather than qualitative attributes. We examine the scheduling of time-critical workflows. The tasks in a workflow are attributed with task durations, resulting in the metagraph analogue of a PERT/CPM network. However, the resulting critical path calculations are richer and more complex than has been the case with the directed graphs used in PERT/CPM project networks.

We will examine these five issues in the five sections below.

1.   REPRESENTING WORKFLOWS AND PROCESSES WITH METAGRAPHS

Most organizations are organized by function, such as purchasing, manufacturing, marketing, engineering, and accounting. Human, physical, and financial resources are often managed by function, and most coordination takes place within functions rather than between functions. Because of this, these functions are often called "stovepipes", stressing the fact that most communication and coordination takes place vertically (up and down within each stovepipe), rather than horizontally (across stovepipes). But many organizations are finding that they must also manage processes that cross functional (or stovepipe) boundaries, such as order fulfillment and new product introduction. These processes not only span the organization's separate functional units but also integrate the organization with other organizations – for example, in supply chain management.

The problem is that although organizations are managed by function, processes are the entities that deliver value to the customer (Davenport, 1993; Marschak, 1995; Thomson 1995). In addition, some of these processes are becoming quite complex, in terms of their size and the structure of both the tasks and information flows that make up the processes. To some degree this size and complexity can be addressed intuitively – that is, by thinking long and hard about the processes. But many people are turning to the development of formal models of business processes (Barua, Lee and Whinston, 1996).

Another concept similar to that of a process is that of a workflow. The difference is that a process may contain Boolean information elements – for example, whether a loan request exceeds a certain amount. A workflow is an instantiation of a process for a set of particular values of the Boolean information elements. In this case the process could result in either of two workflows, one in which the loan request exceeds the target amount and one in which it does not. Thinking ahead for a moment, we might anticipate that processes will be modeled as conditional metagraphs, and workflows will be represented by (unconditional) metagraphs that are obtained by assigning Boolean truth values to the assumptions in the conditional metagraphs.

Who will be involved in the management of processes and their resulting workflows? As it turns out, there will be several such persons (Hammer and Champy 1993; Khoshafian and Buckiewicz 1995; Hammer 1996), and these different people will need to understand different aspects of these processes and workflows. Senior executives will need to understand what information elements are required and produced by a process and what resources are needed. Process managers will need to understand how the tasks in each workflow interact with each other through the information elements they use and produce. In other words, they must understand the flow of information elements through a process and therefore through its resulting workflows. Information technology managers will need to understand how information elements, tasks, and resources interact, so that effective operational and decision support systems can be designed. This suggests that we model processes in a way that facilitates identification and analysis of their associated workflows.

Processes and their workflows can be modeled in several ways using different tools. Four perspectives commonly used in process representation are as follows (Curtis, Kellner and Over 1992; Kwan and Balasubramanian 1997):

1. *Informational modeling* focuses on the informational entities involved in the process, the structure of these entities and their interrelationships. Thus, informational modeling is concerned with the pure inputs to a process, the intermediate information elements in the process, and the information elements that make up the outputs of a process.

2. *Functional modeling* focuses on what tasks are being performed and what informational elements are involved in these tasks. Thus, functional modeling is concerned with the relationships among the various tasks in a process as determined by information elements that are outputs of some tasks and inputs of other tasks.

3. *Organizational modeling* focuses on the agents/resources that will be involved in each task, where information entities are to be stored, and the communication needed between agents/resources. Thus, organizational modeling is concerned with the people, hardware, and software needed for a task to participate in a process. We note that people may be individuals or categories of individuals (e.g., systems programmers). This distinction between individuals may also apply to hardware, software, and any other agents/resources.

4. *Transactional modeling* (also called behavioral modeling (Curtis, Kellner and Over, 1992)) examines issues of timing (sequencing) and control, both within and between the tasks involved in the process. Thus, transactional modeling is concerned with the order in which the tasks are to be performed (some serially and possibly some in parallel), conditioned on the assumptions associated with the tasks and the informational outputs of previous tasks.

Traditionally, these perspectives have been implemented separately. However, we suggest that metagraphs provide a useful and comprehensive foundation for modeling and integrating these perspectives. That is, a significant contribution of metagraphs is that they integrate the informational, functional, and organizational perspectives within a single model. This allows not only graphical visualization of processes, but also their formal analysis, where the analysis will be accomplished by means of an algebraic representation of the graphical structure. In the metagraph view informational entities will correspond to the elements in a generating set, and tasks will correspond to the edges in the metagraph. This construct extends the features offered by traditional graph structures such as digraphs and hypergraphs and allows us to address questions such as the following using algebraic operations on metagraph representations of processes:

1. How do information elements relate to each other through the tasks that use and produce them – for example, which information elements are needed to determine other information elements, and which information elements are intermediate elements in a process that calculates other elements (*informational modeling*)?

2. How do tasks relate to each other through the information elements that they use and produce – for example, if a task is disabled, what other tasks cannot be executed (*functional modeling*)?

3. How do the resources needed to perform various tasks relate to each other through the tasks that use them and the information used and produced by these tasks – for example, what information passes from one resource to another and if a resource is unavailable, what other resources are affected (*organizational modeling*)?

In addition to the questions pertaining to each perspective, there are other questions that span several perspectives. For instance, if a particular resource were to become unavailable, then several tasks might be disabled. How would the workflows in the process be affected? A significant aspect of the integration enabled in our approach is that such questions can also be addressed in a structured manner. In addition, we will address certain transactional issues of timing and scheduling here, task duration and temporal constraints as additional metagraph edge attributes. The result will be a metagraph analogue of the directed graphs used in PERT/CPM analyses.

Essentially, we need a theoretical framework for the representation, analysis, and manipulation of workflow systems. Metagraphs allow different components of processes to be represented both graphically and analytically. This framework also slows us to analyze both connectivity and component interaction of workflows using a single representational construct.

We illustrate this framework with the example of a loan evaluation process. The input to the process is a document containing certain information items describing the applicant and certain characteristics of the loan being requested. Various tasks are used to analyze the application, and possibly to request that additional information be made available, and then to arrive at a decision. Human and computer resources, such as loan officers, loan managers, fax machines, and workstations are used to accomplish these tasks. It is then necessary to determine the flow of information items, the scheduling of tasks, and the allocation of resources. These are often specified in procedures manuals and/or in automated workflow systems.

We now define the central concepts of workflow systems and present some of the questions pertinent to workflow analysis. We begin with the following terms:

1. An *information element* is an atomic data item (e.g., a number, a character string, an image, or an icon) or a collection of atomic data items (as in a document).
2. A *report* is a collection of information elements.
3. A *task* is an ordered pair of reports, the first of which is an input to the task and the second of which is its output. A task is *executed* when the inputs are used to determine the output.
4. A *workflow system* is a set of information elements and a set of tasks, such that the inputs and outputs of the tasks are all in the set of information elements.
5. An *assumption* is a proposition (which may be true or false) associated with a task, such that the assumption must be true for the task to be executed. For example, it may be assumed that the dollar value of a transaction is less than a certain amount.
6. A *resource* is an entity associated with one or more tasks, and the resource must be available if the tasks are to be executed. Several resources may be associated with a single task, and vice versa. Resources may be people, workstations, categories of people (i.e., roles – such as programmer or file clerk), etc.
7. A *process* is a set of tasks that connects one set of information elements, called the *source*, to another set of information elements, called the *target*. All of the inputs for any task in the process must be either in the source or in the output of some other task(s) in the process.
8. A *workflow* is a particular instantiation of a process. Since a process may include decision points that can cause the process to branch in different ways during execution, a process can be instantiated into several possible workflows, each one corresponding to a particular set of values for all relevant branching conditions.

The purpose of constructing a framework for workflow management is that it allows us to formulate questions about the relationships among the three important components of workflows: information, tasks, and resources. Nine such questions are shown in Table 9.1, and are answered in the following sections of this chapter.

In order to answer these questions we need an analytical framework that allows us to: (1) capture all of the important elements in the workflow process, and (2) address these questions by means of rigorous analytical procedures rather than visual inspection and intuition. The theory of metagraphs provides a basis for such a framework. In the next section, we summarize the major features of metagraphs and present some extensions to the theory that are pertinent to workflow analysis. Then in the following section, we will use these features to address questions like those listed above, for workflow management.

Table 9.1. Relevant questions about process components during process analysis

Process component	Questions
Information elements	
	1. Given two information elements, is one of them needed to determine the value of the other? Is it needed only under certain conditions and if so, what are the conditions?
	2. Given two sets of information elements, is it possible to determine the value of the second set from the elements in the first set? If not, are there any additional information elements that would make it possible to do so?
	3. Given a complex process, are there any ways to focus on only important information elements, hiding intermediate elements that are needed only to calculate the important ones?
Tasks	4. Given a task that we wish to execute, what other tasks must be executed in order to provide the information needed to do so?
	5. If a task is disabled, what other tasks will be affected – that is, what other tasks cannot be executed?
Resources	6. Given a set of resources, what information passes among them as the tasks that utilize them are executed?
	7. If a resource is unavailable, what other resources are affected – that is, what other resources will be idle because their tasks cannot be executed?
Interactions among components	8. If a resource used in a process is unavailable, which workflows within the process can still be completed?
	9. If an information element is found to be inaccurate, which resources were used, directly or indirectly, in the calculation of that element?

2. VIEWS OF WORKFLOWS

Each information element in a workflow can be represented as an element of the generating set X, or more specifically, of X_v (i.e., the variables in the workflows). There are other elements in the generating set, specifically X_p (the propositions that denote assumptions) and X_r (statements as to the availability of resources), but they are not information elements in the workflow. A collection of information elements comprising a report can then be represented as a vertex, either an invertex or an outvertex. This presumes that each report is either the input or output of some task, which is reasonable for a "black box" analysis of interactions among tasks, as discussed below. Each task is itself represented as an edge in the metagraph. We assume that the input to each task (invertex) is a report, as is its output (outvertex). This assumption is reasonable, since the report comprising each task's input can be composed of elements from one or more reports (and/or manual inputs from some resource).

It follows, then, that each process can be represented by a metagraph. More generally, a metagraph can be used to represent the tasks comprising a collection of possibly related (or overlapping) workflows comprising the process. For example, the risk exposure of a bank, which is determined by a series of steps including both computations of internal data such as outstanding loans, as well as market conditions and other external information, may be used both in the loan evaluation process as well as in planning the bank's insurance coverage. This last point is significant, since most other graph constructs do not allow such overlapped representation. The collective representation of multiple workflows in a single metagraph enables analysis and possible redesign of these workflows in a more comprehensive manner.

Metagraphs, like other modeling techniques based on graph theory, provide a black box representation of reality. That is, a task, as represented by a metagraph edge, is viewed as a pair of inputs and outputs. We are not concerned with happenings inside the black box. Thus, representation of a task by means of a simple metagraph (i.e., one without any additional attributes on vertices or edges) does not take into consideration its duration, the quality of work done, the value added by the task to any process, and the type of monitoring and control needed to detect errors in the task. However, some of this can be included in the representation by attaching attributes to the edges. We will see an example of this in Section 5 of this chapter, where we will include consideration of time, and specifically task duration, in a metagraph.

Although metagraphs are limited in this regard, they do have a powerful advantage. First, metagraphs model the essential structure of a workflow system, in that they allow for an explicit representation of the components of the system and the interactions among them. Thus, they can be used to determine what sorts of information can be furnished by a workflow system, given the information processing capabilities of its components. Second, as we will see, metagraphs allow for multiple views of a workflow system – an element-centric view, a task-centric view, and a resource-centric view. Since each of these views are themselves metagraphs, the same mathematical machinery can be used for all of them. Third, it is possible that metagraphs can be extended to include some of the contents of the black box mentioned above – for example, by labeling the edges with estimates of costs, error rates, etc. However, such an extension is beyond the scope of this chapter and this book.

To complete the representation, a workflow can be represented as a metapath from a set of information elements comprising a source to another set comprising the target. Assumptions underlying each task can also be represented in the metagraph, by augmenting the generating set with a set X_p of propositions and including the relevant propositions in the invertices of the task edges. And finally, resources needed for each task can also be represented, by further augmenting the generating set with a set X_r of resources. Then, the resources required in each task can be represented as additional inputs to the corresponding edge. Note that the separation of the generating set into the three component sets X_v, X_p, X_r is not done merely for convenience. The primary motivation for this separation is that the evaluation of elements from each set is different. While information elements can have any value from their particular domain, propositions evaluate to either "true" or "false" (with a task becoming viable only if all its assumptions evaluate to "true") and resources evaluating to either "available" or "unavailable" (with a task becoming viable only if all its resources are available). From a visualization perspective, the assumptions underlying a task and the resources it needs can be presented to the user as labels on the edge itself, rather than as invertex assumptions. However, the invertex representation may be more intuitive.

Consider the example illustrated in Figure 9.1, illustrating a workflow process that determines whether an application for a property loan is to be accepted or rejected. The workflow process is modeled as a conditional metagraph, with the incidence matrix shown in Figure 9.2. The information elements in the workflow process, represented by elements in the conditional metagraph, are as follows:

- *AC*: account data relevant to the applicant;
- *APD*: data about the applicant contained in the application;
- *CH*: credit history of the applicant;
- *PD*: data about the property for which the loan is being sought;
- *CD*: data about comparable properties;
- *CR*: applicant's credit rating;
- *AV*: the appraised value of the property;
- *LA*: the amount of the loan;
- *BP*: the current bank portfolio of loans;
- *LR*: the level of risk associated with the loan;
- *RE*: the bank's current risk exposure;
- *YES*: a statement that the application is approved;
- *BR*: a statement that the loan being applied for is a bad risk;
- *NO*: a statement that the application is rejected.

The tasks in the workflow depend not only on the information elements identified above, but also on the following assumptions:

- *AR*: whether the level of risk is acceptable;
- *MR*: whether the loan application represents a marginally bad risk;
- There are eleven tasks in the workflow process, as follows:
    - e_1: the branch manager uses account data and applicant data to calculate the applicant's credit rating;

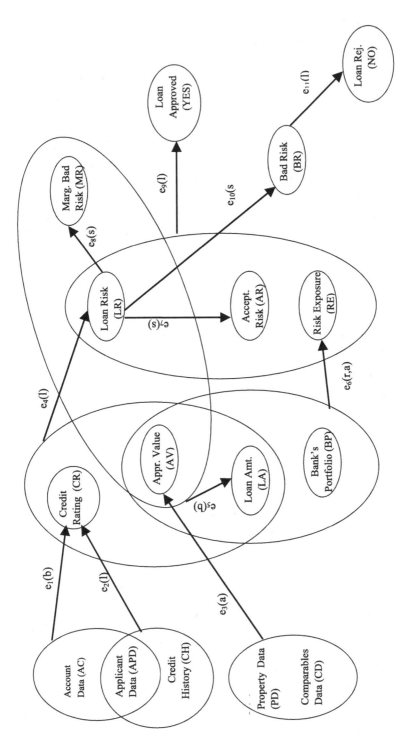

Figure 9.1. A loan process metagraph. Edges (tasks): e_1 – credit rating process; e_2 – alt. CR process; e_3 – property appraisal; e_4 – risk assessment; e_5 – loan amt. reduction; e_6 – risk exposure assessment; e_7 – acceptable risk assessment; e_8 – marginal risk assessment; e_9 – loan approval; e_{10} – bad risk assessment; e_{11} – loan rejection. Resources: l – loan officer; a – appraiser; r – risk analyst; b – branch manager; s – system.

G	e_1	e_2	e_3	e_4	e_5	e_6	e_7	e_8	e_9	e_{10}	e_{11}
AC	-1										
APD	-1	-1									
CH		-1									
PD			-1								
CD			-1								
BP						-1					
YES									+1		
NO											+1
CR	+1	+1		-1							
LA				-1	+1	-1					
AV			+1	-1	-1	-1					
LR				+1	-1		-1	-1	-1	-1	
AR							+1		-1		
RE						+1			-1		
MR				-1				+1			
BR										+1	-1
b	-1			-1							
l		-1		-1					-1		-1
a			-1		-1						
r					-1						
s							-1	-1		-1	

Figure 9.2. The incidence matrix for the loan evaluation metagraph in Figure 9.1.

e_2: the loan officer uses applicant data and the applicant's credit history to calculate the applicant's credit rating;

e_3: the property appraiser uses data about the property along with data about comparable properties to calculate the appraised value of the property;

e_4: the loan officer uses the applicant's credit rating, the appraised value of the property, and the loan amount to calculate the level of risk associated with the loan;

e_5: if the risk of the loan is determined to be a bad risk, the branch manager uses the appraised value of the property and the level of risk associated with the loan to calculate a new loan amount;

e_6: the risk analyst and the property appraiser use the appraised value of the property, the loan amount, and the current bank portfolio of loans to calculate the bank's current risk exposure;

e_7: the system examines the risk associated with the loan and performs the calculations needed to determine whether the risk is acceptable;

e_8: the system examines the risk associated with the loan and performs the calculations needed to determine whether the risk is a marginally bad risk;

e_9: if the level of risk is acceptable, the loan officer uses the risk associated with the loan and the bank's risk exposure to decide whether
to approve the loan;

e_{10}: the system examines the risk associated with the loan and performs
the calculations needed to determine whether the whether the loan
application represents a bad risk;

e_{11}: if the loan application represents a bad risk, the loan officer rejects
the application.

Finally, there are five resources used in the process, and these are shown
in parentheses on the edge label (although semantically, they are treated as
additional invertex elements), as follows:

- a: a property appraiser;
- b: the branch manager;
- l: a loan officer;
- r: a risk analyst;
- s: an automated system.

3.   ANALYSIS OF INFORMATION INTERACTIONS

We now show how the properties of metagraphs can be applied to the analysis of interactions among information elements. The analysis is operational-
ized through the use of the A and A^* matrices.

First, the role of each information element x can be analyzed by examining the row and column corresponding to x in the matrices. For instance, each
triple in the row corresponding to x in A^* represents a simple path from x to
some element, and identifies the coinputs and cooutputs of the path. Similarly,
each triple in the column corresponding to x represents a path from some element to x. We can also perform a number of analyses using this information,
such as the identification of necessary coinputs or tasks between any pair of
elements x and y (coinputs and edges that appear in each triple in the corresponding cell in the A^* matrix), and the identification of any cycles through
x (identified by $a_{ij}^* \neq \varnothing$). For our example in Figure 9.1, we may want to
know if it is necessary to have the account data (AC) in order to determine the
bank's risk exposure (RE). The visual representation may suggest that there is
no connection between these elements. However, the cell in the A^* matrix corresponding to the row for account data (AC) and the column for risk exposure
(RE), has the value $a_{AC,RE} = \{\langle\{AV\}, \{CR, LA, LR, MR\}, \langle e_1, e_4, e_8, e_5, e_6\rangle\rangle\}$,
indicating that if the appraisal value (AV) as well as the account data is known,
then the risk exposure can be computed using the tasks corresponding to the
list $\langle e_1, e_4, e_8, e_5, e_6\rangle$, which would also yield the applicant's credit rating
(CR), the loan amount (LA), the loan risk (LR) and whether the case involves

marginally bad risk (*MR*). (We do not show the entire matrix here, only relevant cells are described as needed.) This shows that indeed *AC* does affect *RE*. Examination of the cooutput also reveals that *MR* is one of the intermediate outputs. This indicates that the dependence of *RE* on *AC* occurs only when the case is evaluated as marginally risky. Such analysis can facilitate scheduling tasks in workflows, as well as resource allocation.

Since a process is composed of a set of tasks that connect a source set of information elements to a target set of information elements, each workflow can be represented by a metapath from the underlying process's source to its target. This implies that a variety of metapath analysis mechanisms can be applied to workflow analysis. For example, given a source and target, the A^* matrix can be used to identify the possible workflows for the process. In other words, a metapath search can help us to determine whether a process is functionally complete or not. That is, if there is no metapath available from the process source to the process target, then additional tasks will have to be included, or else some or all of the component tasks will have to be redesigned. For instance, we might assume that given all the information in a completed loan application, such as account data (*AC*), applicant data (*APD*), credit history (*CH*) and property data (*PD*), the loan process could be completed (i.e., the values of the loan amount (*LA*) and *YES* could be computed). However, if we try to construct a metapath from {*AC*, *APD*, *CH*, *PD*} to {*LA*, *YES*}, we would fail. Based on visual inspection, we might add the data on comparable properties (*CD*), and try again. However, even then, no metapath is found. The element *BP*, representing information about the bank's existing loan portfolio, persists as a coinput. This analysis indicates the need for *BP* as an essential input for successful loan evaluation. On the other hand, there is a metapath from the application data {*AC*, *APD*, *CH*, *PD*, *CD*} to {*NO*}, indicating that a complete workflow for unacceptable cases can be completed without information about *BP*. Again, this conclusion can be useful in designing the process (e.g., *BP* is acquired only after the loan data leads to instantiation of acceptable risk (*AR*)). Furthermore, if there are multiple workflows possible for a given source and target, the concept of a dominant metapath can be applied to facilitate choice among them. Conversely, given a specific workflow, the same property can be used to determine whether this workflow is efficient (i.e., whether it corresponds to a dominant metapath), or whether some alternative workflow may be preferable. Again, in the loan process, the loan amount adjustment task is superfluous, unless the application reflects marginally bad risk (i.e., *MR* is evaluated as True and *AR* is not). Additional analyses include identification of source-dominant metapaths for a given target set (i.e., metapaths that require a minimal source set for the given target set, even if each such metapath includes additional edges). The role of the applicant's credit history *(CH)* in the workflow (metapath $\{e_1, e_2, e_3, e_4, e_6, e_7, e_9\}$) to loan acceptance *(YES)* exemplifies this case. Initially, we might assume that the source for this workflow is $\{AC, APD, BP, CH, CD, LA, PD\}$, but then discover that this metapath is dominated by the metapath $\{e_1, e_3, e_4, e_6, e_7, e_9\}$, with source $\{AC, APD, BP, CD, LA, PD\}$, which does not include *CH*.

The projection operation on a metagraph is another useful construct for analyzing workflows. By focusing on a specific subset of information elements $X'_v \subset X_v$ it displays the relationships among these elements by identifying the processes that relate these elements. From a visualization standpoint, the projection view of a workflow metagraph is valuable, since it focuses attention on a few important elements and tasks. At the same time, the analytical structure underlying the view, in terms of the composition of each projected edge, enables identification of the structure of specific tasks making up the projection. For instance, a projection over appraised value *(AV)*, bank portfolio *(BP)*, loan risk *(LR)* and *YES* would show an edge from $\{AV, BP, LR\}$ to $\{YES\}$ (propositions such as *AR* and *MR* can be hidden in projections, to simplify visualization), which illustrates that under some conditions, the factors determining loan acceptance are the appraised value of the property, the loan risk and the bank's current loan portfolio. This is difficult to ascertain visually from the detailed metagraph, yet is obvious in the projection.

Another abstraction that proves valuable in workflow analysis is a context. Recall that the assumptions applicable to each task in a workflow can be included as propositions in the metagraph representation. Given a set T of known true propositions and a set F of known false propositions, the applicable workflows under these conditions can be identified by constructing the corresponding context of the metagraph. The context metagraph can then be analyzed in all the ways described above. As with the projection operation, the context metagraph provides a means for simplifying complex workflows and focusing attention to relevant tasks and elements. Contexts also help process designers avoid unpleasant situations. For example, given a process defined by a specific source and target, verifying functional completeness of the process in various relevant contexts (by ensuring the existence of relevant metapaths in each context) can help avoid nasty surprises such as the process failing under certain conditions. In our example, we can construct contexts for the different levels of loan risk represented by the propositions *AR*, *MR* and *BR*, and examine the three alternative workflows that result for the loan evaluation process.

Thus far, we have answered questions 1, 2, and 3 posed in Table 9.1. In addition, other connectivity properties, such as cycles and bridges, can be analyzed using our algebraic approach. For example, the cycle through the edges $\{e_4, e_5\}$ can be identified using the diagonal elements of the A^* matrix, and the fact that the edge e_4 is a bridge from the application data to either loan outcome can be algorithmically determined. Thus, any software for workflow analysis based on metagraphs would have a user-friendly GUI for purposes of visualization, but its operation would use structured procedures based on algebraic representations of metagraphs which could be used to answer questions such as those we have discussed in this section.

4. ANALYSIS OF TASK INTERACTIONS

We now turn to the analysis of tasks in workflows. Recall that tasks are represented in the metagraph as edges, and they appear as the third component of each triple in the adjacency matrix of the metagraph.

In workflow analysis, a number of questions about tasks and their role can arise, as exemplified by the questions listed at the beginning of this chapter. For such analysis, it is useful, both from visualization and analytical viewpoints, to have a task-centric view of the workflow system, as opposed to the element-centric view considered so far. Put another way, it would be useful to have tasks as elements of the generating set, and edges linking sets of related tasks. This can be achieved using a simplified version of the inverse metagraph. In the context of workflow metagraphs, we call the resulting construct the *task interaction metagraph* (TIM) for the workflow system.

To construct the task interaction metagraph using Procedure Inverse in Chapter 4, we modify the procedure as follows:

1. Step 3 of the procedure is not used, since we only represent interactions among tasks in the TIM. Thus, pure inputs and pure outputs are excluded.
2. Edge labels are simplified to specify only the information elements.

This is illustrated in Figure 9.3 for our example. The TIM has as its elements the tasks (edges) of the original metagraph, and each edge represents a situation in which one or more tasks (in its invertex) communicates with one or more tasks (in its outvertex) by providing information to them. In Figure 9.3, the edge from e_4 to e_7, e_8, e_{10} illustrates the dependence of the latter three tasks upon e_4 for the value of the loan risk (LR). From the closure of the TIM's adjacency matrix, we can identify the metapath $\{LR, BR\}$ from e_4 to e_{11}, which shows that once e_4 is executed, the edges in the metapath can be executed without interaction with any other tasks. On the other hand, the lack of a metapath from e_4 to e_9 indicates that other tasks besides e_4 have to provide information in order to execute e_9. However, there is a metapath from $\{e_1, e_3, e_5\}$ to $\{e_9\}$. This shows that the edges in that metapath can be used to enable e_9, without interaction with any other tasks. Similarly, the cycle through the edges e_4 and e_5 reveals that these tasks may be executed multiple times in workflows (metapaths) in which they both appear. Such analysis, as exemplified by questions 1 and 2 in Table 9.1, can help in designing workflows, since

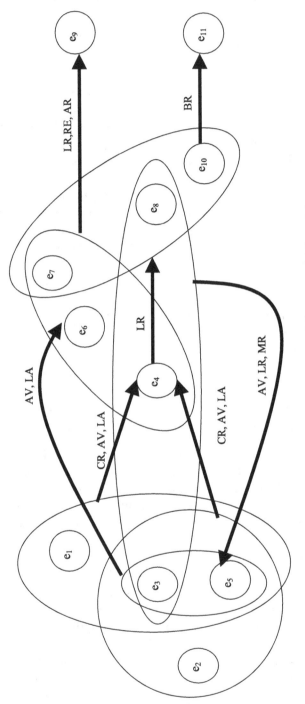

Figure 9.3. Task interaction metagraph for process in Figure 9.1.

coordination of tasks is only necessary when there are dependencies among them.

The TIM can also be useful in analyzing the impact of one or more tasks failing during a process. Another related question is whether a particular task is essential for a workflow. Since a task could produce multiple outputs, each of which is used in possibly several other tasks, such questions cannot always be answered simply through the visual representation of either the process metagraph or its TIM. However, by testing whether a given workflow meta-path is dominant or not from the closure matrices for either representation, these questions can be answered in a structured manner. Further, if we want to know whether a particular task e_{pi} in a workflow $\{e_{p1}, e_{p2}, \ldots, e_{pn}\}$ would be disabled by the failure of another task e_j, the closure of the TIM's adjacency matrix can be utilized. If all the paths in the column corresponding to e_j and rows $\{e_{p1}, e_{p2}, \ldots, e_{pn}\}$ contain e_j, then indeed e_j is essential for e_{pi}. For example, in the TIM in Figure 9.3, every path from $\{e_1, e_3, e_5\}$ to e_9 contains e_4, which is thus essential for the workflow from $\{e_1, e_3, e_5\}$ to $\{e_9\}$.

We have now answered questions 4 and 5 posed in Table 9.1. The strength of metagraph-based task analysis in workflow modeling is that we can perform both visual inspection and algebraic analysis of workflow tasks using the same operations and methods as those we used earlier for analysis of information elements and their interactions. In other words, the approach integrates functional and informational analysis of a system (Curtis, Kellner and Over, 1992), which is difficult using traditional tools for process analysis.

5. ANALYSIS OF RESOURCE INTERACTIONS

We now turn to the analysis of the role of each resource in the workflow, and interactions between different resources. A resource may be human (a person, group, team or task force), or equipment (e.g., computers, software packages or programs, and databases). Furthermore, a resource may be a particular entity (e.g., a specific person) or a class (e.g., a functional role, such as a financial analyst). There is a many-to-many mapping between tasks and resources – each task may require several resources, and each resource may be required in several tasks.

An important part of business process design is the effective allocation of resources to tasks. There are two dimensions to this allocation problem. The first is the functional interaction among different resources, the tasks they perform and the information elements they use and produce. These considerations and their analysis impact the design of processes and their workflows. The second dimension is the temporal constraints applicable to the interactions that impact the operational control of workflows during their execution. This is important for monitoring and control of workflow execution (at run time), in addition to

R	b	l	a	r	s
CR	e_1	$e_2, -e_4$	0	0	0
LA	e_5	$-e_4$	$-e_6$	$-e_6$	0
AV	$-e_5$	$-e_4$	$e_3, -e_6$	$-e_6$	0
LR	$-e_5$	$e_4, -e_9$	0	0	$-e_7, -e_8, -e_{10}$
AR	0	$-e_9$	0	0	e_7
RE	0	$-e_9$	e_6	e_6	0
MR	$-e_5$	0	0	0	e_8
BR	0	$-e_{11}$	0	0	e_{10}

Figure 9.4. The R matrix for the loan evaluation process.

workflow design. The approach presented in this chapter addresses only the first dimension, since we do not include temporal attributes for tasks (although our approach can be extended towards this end with attributed metagraphs). Thus, our focus is on understanding how different resources interact with each other, the tasks through which these interactions occur, and the information that they exchange.

Interactions among resources can be specified by a *resource interaction metagraph* (RIM), which shows where resources provide information to each other through a sequence of two successive tasks. This is accomplished by using the element flow metagraph. If the element set X' for the element flow metagraph is restricted to the set of resources, then the result is in fact the RIM. For our example, the G_1 and G_2 matrices for Procedure EFM correspond to the sets of rows in Figure 9.2 denoted as G_1 and G_2. The R matrix that results from the first step is as in Figure 9.4, and the result of the procedure is the RIM illustrated in Figure 9.5.

Using this metagraph we can determine which resources depend on other resources to provide information. That is, we can identify cases in which a resource is used to perform a task that provides information to another which uses the second resource. For example, for the loan evaluation process, the branch manager interacts with the loan officer by providing the latter with the applicant's credit rating for the risk assessment task;. the loan officer in turn performs the risk assessment and returns the risk value of the loan to the branch manager. In addition, when the branch manager performs the loan amount reduction task, she also provides a loan amount to the loan officer for the loan risk assessment. This interaction can be easily visualized from the RIM, but is difficult to visualize from the original metagraph.

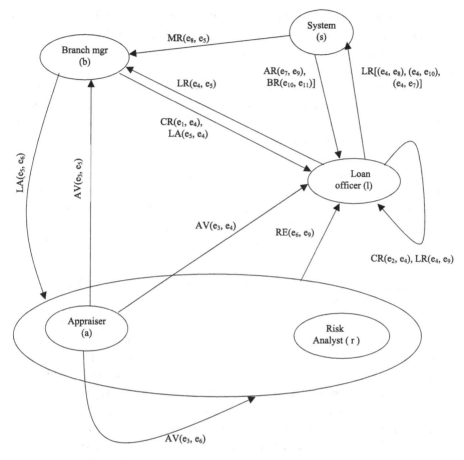

Figure 9.5. Resource interaction metagraph for loan process.

Another application of the RIM is to analyze the impact of resource failure. For instance, we may want to examine the impact of a failure in the automated system (s). From Figure 9.5, it is clear that this directly impacts only the loan officer and the branch manager, who directly interact with the system. In other words, the tasks performed by the appraiser and the risk analyst can still be performed. However, since at some point, there may be indirect dependencies between the system and these resources, it is important to identify the indirect dependencies as well. Since the RIM is a metagraph, it has an adjacency matrix and its closure, and these can be used to identify the relevant dependencies. Thus, if the row in the closure matrix corresponding to the system has any entries in the columns for the appraiser and the risk analyst, then there is a dependency, and at some point, these resources will be impacted by a failure in the system. This is indeed the case, as indicated for instance by the path from s to $\{a, r\}$ through b (the information flows $MR(e_8, e_5), LA(e_5, e_6)$). There are other paths too, such as the path through l and b (the information flows $AR(e_7, e_9)$, $BR(e_{10}, e_{11})$), and another involving the loop through l. Once these dependencies are identified, their roles can be analyzed using either the original metagraph or the RIM.

In this section, we have presented a third view of workflow metagraphs, the resource interaction metagraph. This view (i.e., the RIM), the resource task assignment matrix, and the transformation operation used to generate it from the original workflow metagraph are new metagraph constructs. Using this view, we can answer a variety of important questions about the role of resources in the workflow and its underlying process, including questions 6 and 7 posed in Table 9.1. As with the analysis of tasks discussed in Section 3 above, while the original workflow metagraph can be used to address such questions, the RIM significantly enhances both visualization as well as analysis of resource interactions, which in turn can lead to better workflows and process design.

6.   INTERACTIONS AMONG DIFFERENT TYPES OF COMPONENTS

The previous subsections focused on questions about the interactions among components of the same type – that is, among information elements, tasks and resources respectively. We now examine an important additional dimension, namely questions that span different component types.

For instance, we may want to determine the effect of a particular resource's unavailability upon the feasibility of one or more workflows. In the loan process, if the risk analyst (r) were unavailable, could the loan evaluation still be completed? We answer this question by restricting the metagraph in Figure 9.5 to the context where r is unavailable. This results in edge e_6 being disabled. If we now try to construct metapaths to $\{YES\}$ and $\{NO\}$ respectively from $\{AC, APD, P\acute{D}, CH, BP\}$, we find that we can still complete the workflow for rejected loans, but not for approved loans. Thus, the risk analyst is essential for making positive loan decisions, but not for negative ones.

On the other hand, if the automated system resource (s) were unavailable, then the edges e_7, e_8 and e_9 would be disabled, which would disconnect the source and target for the process, rendering the process infeasible. This is important, since it indicates to the process designer that either the resource s should be designed to be highly reliable, or else that the process should be enhanced with additional tasks and/or resources that could be used to reduce the criticality of the automated system.

Another question that may arise is – what resources are used to produce a given set of information elements? For example, assume that bank management raised a question about the loan risk assessment values. To determine all the resources that contributed to the valuation of this element, we could examine the process metagraph visually and conclude that the branch manager (b), loan officer (l) and appraiser (a) were the relevant resources. However, by examining the cells a^*_{iLR} in the A^* matrix for all resource rows i and the column LR, we would find that the automated system s is also relevant. Especially in complex processes with many tasks, such analysis can help determine accountability for tasks and outcomes in a systematic manner.

We have now answered questions 8 and 9 posed in Table 9.1.

7. SYNTHESIS OF PROCESSES

While processes are sometimes designed from scratch, there are also many situations where a process has to be constructed from multiple existing processes. This is particularly true when processes are redesigned within an organization, or when multiple organizations merge or reorganize. Thus, an important area of process design is the analysis of processes that are composed of two or more component subprocesses.

In order to appreciate the types of analysis would be relevant to the synthesis of a process from multiple components, one must understand the implications of such synthesis. Since each process has multiple possible workflows, the combination of multiple processes, each with multiple workflows, can lead to a combinatorial explosion of possible workflows through the new synthesized process. Even if each of the workflows in each of the component processes is well-structured, this may not be true of the synthesized workflows. Ensuring that every synthesized workflow is well-structured can thus be a daunting task, and the use of computer-based analysis tools can be of great value.

What are the possible problems that could arise when several well-structured workflows are combined? To see this, consider the metagraph representation of workflows. As discussed earlier, a process can be represented as a metagraph with a single "Start" vertex and a single "End" vertex. If the process is well-structured, then there should be exactly one applicable workflow in each possible interpretation. This means that there should be exactly one corresponding metapath in each interpretation as well. This can be checked by enumerating the possible interpretations, and in each case, to find all applicable metapaths $M(Start, End)$ in the process metagraph. If for some interpretation there is no such metapath, then the workflow corresponding to that interpretation has to be modified and/or augmented with tasks to achieve connectivity. Similarly, if there are multiple metapaths in any interpretation, then appropriate tasks from the process have to be modified and/or removed to eliminate the multi-determinacy.

One way to visualize the synthesis of two processes is as the union of their metagraph representations, with the start and end nodes being redefined as appropriate. This in effect results in superimposing the identical elements and adding the adjacency matrices. An interesting question then is, if each of the component processes are still complete, in that the process can complete in all feasible situations. In terms of the metagraph representation, this amounts to checking whether there is a feasible workflow through the combined metagraph under all relevant interpretations. From Theorem 6.1 of Chapter 6, we know that the union of two metagraphs maintains full connectivity if it does not introduce any cycles. This is a useful result, because it implies that preservation of full connectivity can be verified by checking for acyclicity of the combined metagraph. This in turn is very easy to do, since the closure of the adjacency matrix of an acyclic metagraph would have no elements in the main diagonal.

To illustrate this idea, we can use the example of loan evaluation processes. Consider the Loan Risk Process represented by the metagraph S_1 illustrated in Figure 9.6.

The pure inputs to this process are AC, AP, AV, and LA, and the process has only one pure output, namely LR, while CR is an intermediate element. Since there are no propositions, there is only one interpretation, and there is only one metapath from the pure inputs to the pure output. Thus, the process is fully connected, and also acyclic.

Now consider the Loan Decision Process represented by the metagraph S_2 in Figure 9.7. The pure inputs of this process are PD, CB and LR, and the pure outputs are LD and LA. There are two propositions, $?RL$ and $?RH$, which are related through the constraint R that exactly one of these propositions be true and the other be false. Thus, there are two possible workflows. In each case, the pure inputs are the same. If $?RL$ is true, then the pure output is LD, and there is a single metapath from the pure inputs to the pure output. If $?RL$ is false (and therefore $?RH$ is true), then LA is the pure output, and there is a single metapath from the pure inputs to the pure output. Again, the process is fully connected, and also corresponds to an acyclic metagraph.

What happens if we combine these two processes? The resulting process can be represented by the union of the two metagraphs in Figures 9.6 and 9.7, and is shown in Figure 9.8. Given that both the component metagraphs were fully connected, is this also true of the combined metagraph? Interestingly, this is not the case. To see this, consider the interpretation where $?RH = TRUE$ & $?RL = FALSE$. It turns out that in this interpretation, there is no metapath from the "Start" node of Figure 9.8 to its "End" node.

The reason for this is also interesting, and can be determined by testing the combined metagraph for cyclicity. In this simple example, it is easy to see that there is indeed a cycle in the metagraph, through the elements LA, LR, and $?RH$, and this cycle can be identified by non-empty diagonal members of the closure matrix A^*. Thus, in the above interpretation, the synthesized

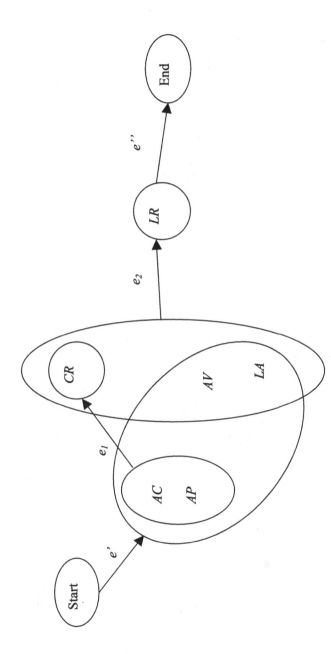

Figure 9.6. Loan risk metagraph S_1: *AC* – account data; *AP* – loan application; *AV* – appraised value; *CR* – credit rating; *LA* – loan amount; *LR* – loan risk.

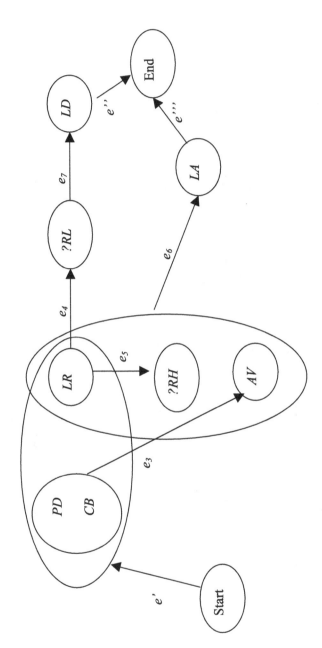

Figure 9.7. Loan decision metagraph S_2: *AV* – appraised value; *CB* – comparables basis; *LA* – loan amount; *LD* – loan decision; *LR* – loan risk; *PD* – property data; *?RH* – risk high?; *?RL* – risk low?

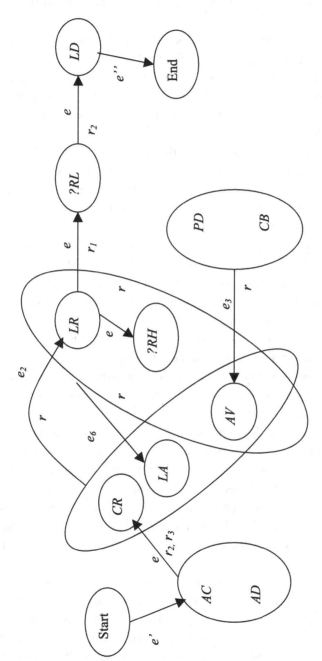

Figure 9.8. The combined metagraph $S_3 = S_1 \cup S_2$: r_1 – risk analyst; r_2 – loan officer; r_3 – loan clerk.

process would fail to complete, even though the component processes would each be able to complete. Also, the cause is a cycle that was introduced by the synthesis itself.

Even this simple example illustrates the potential value of constructing a formal analytical tool for testing the effects of process synthesis. Each of the steps in this analysis can be automated, and thus a metagraph-based tool can alert process designers to potential problems before they are implemented. At the same time, it is important to recognize that the presence of a cycle does not necessarily imply a problem with the process. In our example, if we specify an initial value for *LA* and then iterate through the evaluation process until a value for *LA* is reached for which *?RL = TRUE*, the process ultimately exits the cycle. For example, we can use *PD* to calculate an initial value for *LA*. We then use *LA*, *CR*, and *AV* to calculate *LR*. If the value of *LR* is too high so that *?RH = TRUE*, we reduce *LA* using e_6 (for instance, the task e_6 might reduce the loan amount by a fixed percentage, say 20%), and the revised loan is evaluated. If *?RH = FALSE* (and therefore *?RL = TRUE*), then the process terminates with a loan approval decision. Otherwise, e_6 is used again to further reduce the *LA* and the cycle repeats (the cycle always terminates since the risk of a loan amount near or at zero is implicitly low).

We have learned from this example that the synthesis of well-structured processes can lead to a composite process that is not well-structured because the synthesis operation may introduce a cycle into the composite process. On the other hand, the introduction of cycles need not always prevent the composite process from being well structured.

Another aspect of well-structuredness of processes is non-redundancy. As discussed earlier, a process is non-redundant if it has at most one feasible workflow through it in each feasible interpretation. Non-redundancy is a desirable feature, since it ensures that the process is both deterministic and predictable. In the metagraph representation, this amounts to ensuring that there is at most one metapath from the start to the end node of a process metagraph in each context, which can be verified using structured procedures for finding metapaths.

In the context of process synthesis, an interesting question is whether the synthesis of two non-redundant metagraphs also non-redundant. It turns out that this is always true when the component metagraphs have pairwise disjoint outvertices. Considering again the example represented in Figures 9.6–9.8, we see that indeed, in this case this property is satisfied for the two component metagraphs, implying therefore that the combined metagraph is also non-redundant.

8. DECOMPOSITION OF PROCESSES AND IMPLICATIONS FOR ORGANIZATIONAL DESIGN

Once we have created a synthesized (or aggregated) workflow, using the approach of the previous section, we may wish to decompose this aggregate by removing some of the activities and treating them as separate workflows. For example, we may wish to create a separate management structure for these workflows or to outsource them. There are many managerial and economic reasons for such a decomposition and/or outsourcing, but they are beyond the scope of this book. However, regardless of the motivation, it is important to ensure that any such changes do not disrupt the overall process(es), and we can address this problem using the property of sub-metagraph independence. The result of the decomposition would be that the remaining metagraph would be smaller, and potentially simpler in structure, which in turn would result in a process that is simpler and easier to manage.

Note that it may be useful to extract either a single workflow in some cases, or an entire process in other cases, from a larger containing process. For example, we may wish to outsource risk assessment of a loan in a particular situation (e.g., when the customer does not have an account at the bank); alternatively, it may be appropriate to outsource the risk assessment process as a whole. That is, we may decide to outsource the union of all of the workflows making up the risk assessment process. We assume that a workflow can safely (i.e., without disrupting any other workflow) be extracted from a containing process if the metagraph representing the workflow is independent of the containing metagraph. Then it follows that in order to extract an entire (sub)-process, we would need to consider whether we can extract the union of all of its component workflows.

There are two issues here. The first is whether the union of two independent workflows – that is, two separate workflows that are each independent of the entire aggregate of workflows – will also be independent of the entire aggregate. If that were not the case, then it would be necessary to consider each workflow in turn, and extract it only if it were independent of the current aggregate.

With regard to independence of the union of two independent workflows, we have seen in Chapter 6 (Theorem 6.2) that independence is preserved in this case. In other words, bundling (merging) two independent sub-processes for subsequent decomposition will result in an aggregate independent subprocess. In addition, according to Theorem 6.3 the subprocesses will be independent of each other. We note, however, that redundancy is not necessarily preserved under these conditions. In other words, each workflow might contain proce-

Table 9.2. Independence of submetagraph and RIM

		RIM	
		Independent	Dependent
Submetagraph	Independent	Decomposable organization	Matrix organization
	Dependent	Modular workgroups	Monolithic organization

dures for non-redundantly calculating the same information element, but the resulting union could be redundant.

The second issue is the intersection of two workflows. We may wish to create a new and smaller workflow by pulling out the tasks common to two workflows. If the two workflows are independent, the question is whether independence will be preserved in the intersection. We have seen in Theorem 6.4 that one can create a new workflow by taking all activities common to two independent workflows and decompose the resulting set of common activities. In this case the independence property will be preserved. However, as we have also seen, the property of full connectivity will not necessarily be preserved.

The concept of independence also provides guidelines for process decomposition. Consider a process that is a candidate for decomposition. There are two issues here. The first is whether a workflow, as represented by a submetagraph, is independent of the aggregate process – that is, whether it is an ISMG. The second is whether the collection of resources for this submetagraph, in the form of the corresponding resource interaction metagraphs (RIMs) is also an ISMG of the aggregate RIM. There are four possibilities, as illustrated in Table 9.2.

The first possibility, termed *decomposable organization* in Table 9.2, occurs when both the submetagraph and the RIM are ISMGs. In this case the corresponding workflow is a good candidate for decomposition. Of course, there may be other criteria for treating the decomposed workflow separately. Certain economic and managerial issues should be considered here, as well as traditions and issues of corporate culture. However, this joint independence suggests that there are no structural (or process-specific) impediments to decomposition.

The second possibility, termed *matrix organization* in Table 9.2, occurs when the submetagraph is an ISMG but the RIM is not an ISMG. In this case the tasks are separable but the process resources are shared with other parts of the resource aggregate. This suggests the use of a matrix structure in which the decomposed sets of tasks (corresponding to the ISMGs) are managed by separate project managers. Resource managers would assign the resources centrally to the various projects.

The third possibility, termed *modular workgroups* in Table 9.2, occurs when the submetagraph is not an ISMG but the RIM is an ISMG. In this case the tasks contained in the subprocess use specialized resources but interact with other tasks outside the subprocess. In this case, the resources can be organized in a module (e.g., a workgroup). Thus, even though the tasks performed within the subprocess require interaction in the form of inputs and outputs with other tasks outside the subprocess, the module/workgroup itself requires less coordination with external resources, since only certain resources interact with other resources.

The fourth possibility, termed *monolithic organization* in Table 9.2, occurs when neither the submetagraph nor its RIM is an ISMG. In this case the organization cannot be decomposed for structural reasons. Of course, there may be compelling reasons to decompose because of personalities, organizational culture, economics, geographical locations, traditions, or other reasons. However, process managers should realize that there are structural arguments against decomposition.

We note that these independence conditions must hold for all interpretations of the conditional metagraph representing the process in question. However, in some cases there may be only partial independence – that is, independence in some interpretations but not in others. In this case, process managers and analysts may select the organizational alternatives suggested above for most instances and be prepared to override them in the special circumstances.

We illustrate how an analysis of resource interactions can augment the analysis of metagraph independence for process decomposition, using our loan evaluation example in Figure 9.8. The edge designations appear above the appropriate edge in Figure 9.8, and the resources used in the various tasks are indicated in the figure below the appropriate edge. We can see that there are three resources – r_1: a risk analyst, r_2: a loan officer and r_3: a loan processing clerk. The RIM corresponding to Figure 9.8 is shown in Figure 9.9. The edge labels in Figure 9.9 indicate the nature of the resource interactions. For instance, the label $(CR) \langle e_1, e_2 \rangle$ on the edge from $\{r_2, r_3\}$ to $\{r_1\}$ indicates that the loan officer and loan clerk compute CR in the task e_1 and provide it to the risk analyst for use in edge e_2.

From the RIM, it is apparent that the risk analyst (r_1) always works alone, while the other resources work both alone as well as with each other. Thus, a potential candidate for decomposition is the set of edges using r_1. We can identify these edges from either the RIM (source edges in labels of all edges emanating from r_1), or from the process metagraph. The edges are e_2, e_4 and e_5. What next needs to be examined is whether these three edges form an ISMG.

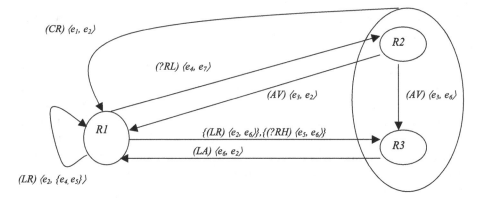

Figure 9.9. The resource interaction metagraph for S_3.

If e_2, e_4 and e_5 are extracted as a separate sub-process, we get the metagraph shown in Figure 9.10 where the extracted sub-process is represented by the new edge e'. Unfortunately, this is not an ISMG, since *LR* violates output dependency. However, this occurs in only one of the two possible interpretations of the process (i.e., when *?RH = TRUE* and therefore, *?RL = FALSE*), and thus represents a case of partial independence. In other words, for all low risk cases (*?RL = TRUE*), the risk assessment sub-process represented by e_2, e_4 and e_5 can be extracted as a separate process, since e_6 does not apply and thus e' corresponds to an ISMG. However, for high-risk cases (*?RH = TRUE*), the fact that *LR* is involved in the sub-process represented by e' is lost, so that S' is not an accurate representation of the workflow.

9.   REPRESENTING TIME-CRITICAL WORKFLOWS WITH ATTRIBUTED METAGRAPHS

It is also possible to attach quantitative (numerical) attributes to metagraph edges. The purpose of this would be to allow certain calculations to be performed. For example, if the attributes are the costs of the tasks represented by the edges, then these attributes can be used to determine the total cost of the tasks appearing in a workflow. If the attributes represent the durations of the tasks, then they can be used to calculate the duration of the workflow in a fashion similar to the PERT/CPM calculations used in project management. If the attributes represent measures of performance, such as degrees of reliability or probabilities of non-failure, then they can be used to determine the performance of the workflow.

These attributes might also be combined. For example, if certain numerical attributes represent both time (i.e., activity durations) and cost, then these attributes might be combined to perform time/cost tradeoffs. If they represent

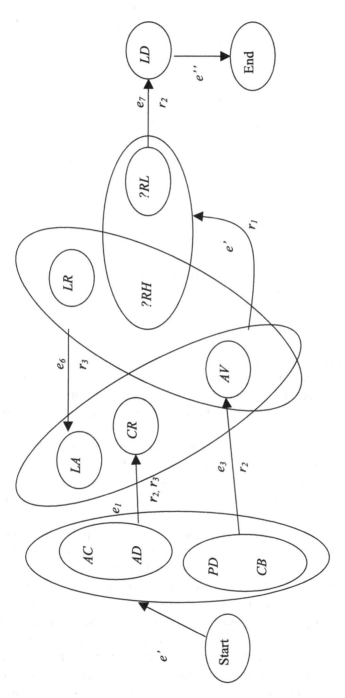

Figure 9.10. The metagraph S'.

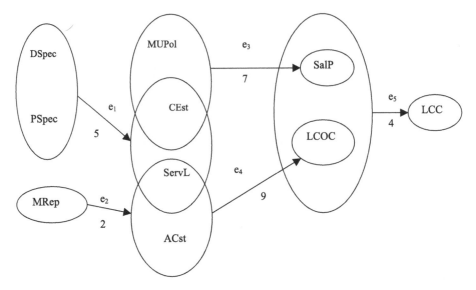

Figure 9.11. Time-constrained workflow metagraph: *ACst* – annual operating cost; *CEst* – mfg. cost estimates; *DSpec* – design specifications; *LCC* – total life cycle cost; *LCOC* – life cycle operating cost; *MRep* – mileage report; *MuPol* – mark-up policy; *PSpec* – production specifications; *SalP* – sales price; *ServL* – service life estimate.

either cost or duration along with probability of non-failure, then they might be used to determine the probability distributions of workflow cost or duration, depending on what will be done if a task represented by an edge should fail. In this chapter we will focus on deterministic activity durations, and we will not consider time/cost tradeoffs.

In this section, we discuss how metagraph-based analysis of workflows can help identify critical activities and time-critical information elements in each workflow. This knowledge can then be used to manage the resources allocated to each activity in the workflow, and ultimately, in the underlying business process.

Consider a workflow metagraph in which each edge is labeled with its expected duration. An example of such a metagraph is shown in Figure 9.11. The generating set consists of ten elements, each describing a document used in a life cycle costing workflow for a vehicle. There are five edges, each representing a task in the workflow. The edge e_1 represents a task that converts design specifications (*DSpec*) and production specifications (*PSpec*) into manufacturing cost estimates (*CEst*) and an estimate of service life (*ServL*). The edge e_2 represents a task that converts the vehicle's mileage report (*MRep*) yet another estimate of service life and also an estimate of annual operating cost (*ACst*). The edge e_3 represents a task that uses the manufacturing cost estimate and the company's markup policy (*MUPol*) to produce a sales price (*SalP*), and the edge e_4 uses the estimates of service life and annual operating cost to calculate the life cycle operating cost (*LCOC*). Finally, e_5 uses the sale price and the life cycle operating cost to calculate the total life cycle cost (*LCC*).

Given the source and target specification for this workflow, it is possible that the actual time available for some of the tasks is greater than the expected time, while other tasks are critical, in the sense that they have to be completed within very tight time constraints. We can identify the tasks and information elements in each category, using analysis similar to PERT/CPM methods in project management (Moder, Phillips and Davis, 1983). However, since the A^* matrix is available for the metagraph representation, it can be exploited in the process.

We treat cases where the outvertices of two or more edges overlap in a manner consistent with traditional project management approaches such as PERT/CPM. That is, when a particular information element is computed by multiple tasks in a workflow, its value is determined only after all these activities have completed. For example, in Figure 9.11, the value of *ServL* can be used as an input to another activity (e.g., activity e_4), only after both activities e_1 and e_2 have been completed. Given any acyclic metapath from a source set B to a target set C critical elements and activities can be found as follows:

Procedure Critical-Path (M, B, C)

Phase 1: Early times

For each element x_i in B, assign the label $Q_i = 0$, and mark x_i as live; let $Q_i = 0$ for all other elements.

Let $E = M(B, C)$

While $E \neq \varnothing$, for each edge e_j in E such that all elements in the invertex of e_j are live, do

Let $T_v = \max_{x_i \in Ve_j}(Q_i)$

For each $x_k \in W_{e_j}$, set $Q_k = \max[Q_k, (T_v + d_j)]$, and mark it as live if it is not already so.

Set $E = E\backslash\{e_j\}$

Repeat

Let $T^e = T^l = \max_{x_i \in X} Q_i$

Phase 2: Late times

Assign $L_i = T^l$ for all elements x_i in C, and mark them as live; set $L_i = \propto$ for all other elements.

Let $E_0 = M(B, C)$

While $E_0 \neq \emptyset$, for each edge e_j in E_0 such that all elements in the outvertex of e_j are live, do

Let $T_w = \min_{x_i \in W_{e_j}}(L_i)$

For each $x_k \in V_{e_j}$, set $L_k = \min[L_k, (T_w - d_j)]$ and mark it as live if it is not already so.

Set $E_0 = E_0 \backslash \{e_j\}$

Repeat

Phase 3: Critical elements

For all $x_i \in X$, if $Q_i = L_i$ then x_i is marked as critical, with completion time $T_i = Q_i = L_i$

END.

DEFINITION 9.1. An invertex V is *critical* if $\max_{x_i \in V}(Q_i) = \min_{x_i \in V}(L_i)$. We note that we will always have $\max_{x_i \in V}(Q_i) \leq \min_{x_i \in V}(L_i)$ as long as Procedure Critical-Path is used to label the metagraph.

THEOREM 9.1. *An invertex V is critical if it contains any critical elements.*

PROOF. Without loss of generality, assume that V contains two elements, a critical element a and a non-critical element b. Let Q_a, L_a, and Q_b, L_b be the early times and late times respectively of these elements. Then

$$\max_{x_i \in V}(Q_i) = \max[Q_a, Q_b], \qquad \min_{x_i \in V}(L_i) = \min[L_a, L_b].$$

Since V is feasible, it follows that $\max[Q_a, Q_b] \leq \min[L_a, L_b]$. However, since a is critical, $Q_a = L_a$, and thus, $\max[Q_a, Q_b] \leq \min[Q_a, L_b]$.

There are three possible cases:

Case 1: $Q_a \leq Q_b \leq L_b$. In this case, $\max[Q_i] = Q_b$, $\min[L_i] = Q_a$ which makes the vertex infeasible unless $Q_b = Q_a$, in which case the vertex is critical.

Case 2: $Q_b \leq Q_a \leq L_b$. In this case, $\max[Q_i] = Q_a$ and $\min[L_i] = Q_a$, and the result follows.

Case 3: $Q_b \leq L_b \leq Q_a$. In this case, $\max[Q_i] = Q_a$ and $\min[L_i] = L_b$, and as in Case 1, the only feasible situation is when the vertex is critical. $\square$

We note that these elements must be contained in a critical invertex, except for those in the final outvertex, in this case LCC.

DEFINITION 9.2. The *slack* in an edge e is defined as

$$slack(e) = \min_{x_i \in W_e} (L_i) - \max_{x_j \in V_e} (Q_j).$$

DEFINITION 9.3. An edge is defined as *critical* if it has no slack.

THEOREM 9.2. *If the invertex of an edge e contains a critical element a with completion time T_a and the outvertex of the edge has a critical element b with completion time T_b such that $T_b - T_a = d_e$ then e is critical.*

PROOF. Since a and b are critical, it follows that $\min_{x_i \in W_e}(L_i) \leq T_b$, $\max_{x_j \in V_e}(Q_j) \geq T_a$.
   Thus, $\min_{x_i \in W_e}(L_i) - \max_{x_j \in V_e}(Q_j) \leq T_b - T_a = d_e$, so that $slack(e) \leq d_e - d_e = 0$, which proves the result. □

THEOREM 9.3. *Each critical edge lies on a critical simple path from some element in B to some element in C.*

PROOF. Since the edges under consideration are all part of a metapath from B to C, each critical edge is on some simple path from an element in B to another in C. Now consider all such simple paths passing through a given critical edge e. If none of these paths is critical, then there is non-zero slack on all of them, which in turn means that e must have non-zero slack, which contradicts the claim that e is critical. Thus, at least one of these paths must have zero slack, and thus is critical. □

   We can illustrate these concepts and how they can be used, using the example workflow metagraph in Figure 9.11. Applying Procedure Critical-Path to this metagraph, we get a labeling of the early and late times for each element as illustrated in Figure 9.12. From Theorem 9.1, we can then infer that the critical vertices are {*DSpec, PSpec*}, {*CEst, ServL*}, {*SalP, LCOC*} and {*LCC*}. The critical edges in the workflow are then e_1, e_4, and e_5 (by Theorem 9.2) and thus the critical path through the workflow is $\langle e_1, e_4, e_5 \rangle$ (by Theorem 9.3).
   Once the critical path is known, scheduling and resource allocation of tasks can be done more effectively. For example, although task e_1 is critical, not all of its outputs are critical – that is, although the service life estimate (*ServL*) must be ready at the earliest possible time (i.e., time $= 5$), the manufacturing cost estimates (*CEst*) can be generated anytime between times 5 and 7. Thus, our analysis helps workflow managers not only with inter-task scheduling but also with intra-task scheduling.
   These results can also be useful in resource allocation. The metagraph analysis enables managers to determine that any resources for tasks e_2 and

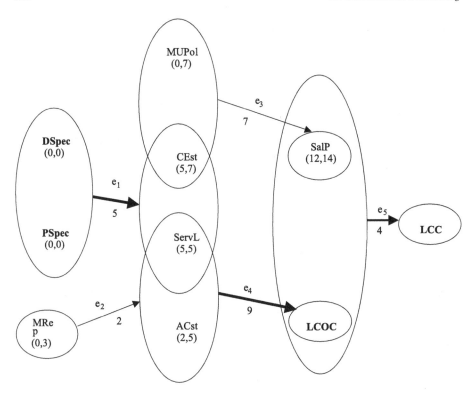

Figure 9.12. The workflow metagraph in Figure 9.11 with critical elements and critical path. The numbers in parentheses beneath each element x_i denote (Q_i, L_i) for that element, the thick arrows denote the critical edges, and boldface denotes critical elements.

e_3 can be reallocated as long as they are not critical – that is, as long as the reduced resources do not extend the task durations more than the existing slacks in the edges. Furthermore, even within critical tasks such as e_1 resources should be allocated to service life estimation with higher priority than to manufacturing cost estimation, unless other factors (e.g., quality or performance) become significant.

Chapter 10

CONCLUSION

We have now completed our presentation of metagraph theory and applications. We began by defining a metagraph as a collection of directed set-to-set mappings, where the sets are subsets of a generating set, at most one of the sets in any edge is null, and for any edge the two sets defining the edge are disjoint. We then developed an algebraic theory of metagraphs and applied it to metagraph connectivity, metagraph transformations (especially projection), assignment of attributes to edges, assumptions and conditional metagraphs, and the properties of sub-metagraphs. Finally we examined the application of metagraphs to the structuring of decision support (i.e., data, model, and rule management) systems and workflow systems.

We conclude in this chapter by addressing three topics. The first is the metagraph modeling process. We center our attention on a previously proposed model development life cycle, similar to the systems development life cycle used in information systems analysis and design, and discuss its application to metagraphs. The second is the construction of a metagraph workbench – a computer-based tool for constructing metagraphs in a variety of contexts. The third topic is the possible application of metagraphs to a quite different area not discussed previously – social networks.

1.   THE METAGRAPH MODELING PROCESS

The metagraph modeling process, like other information systems development processes, is accomplished by means of a set of largely sequential but occasionally overlapping design tasks. These tasks are often called stages. These stages have been studied in detail for general information processing systems in the form of a systems development life cycle or SDLC (Dennis and Wixom, 2000; Hoffer, George and Valacich, 1999). They have also been studied more specifically for a more specialized process, model construction, in the form of a model development life cycle, or MDLC (Blanning, 2003). We explain briefly the MDLC and interpret it in the context of metagraphs.

A life cycle is a sequence of stages that are needed to accomplish a goal. Although the stages are presented as a sequence, it is understood that there will be a certain amount of overlap as the life cycle is implemented, because earlier stages may have to be revisited after later ones have been initiated. Typically, the first stage is a recognition that problem exists and requires the initiation of the remainder of the cycle, and the last stage assumes that a system has been designed and implemented and must now be maintained and possibly modified. The intermediate stages may include feasibility study, systems analysis, and systems design.

The life cycle concept is of special interest because it helps to structure our thinking about the underlying processes and it often provides a framework for preparing progress reports. Very often these stages require a formal statement or signoff by the customers of the system, and thus provide a formal channel of communication between designers and users. This is of special importance because it is believed that the principal cause of system failure is lack of user acceptance, and that this in turn is the result of poor communication between designers and users. In addition, mistakes made during the early stages may not become manifest until the later stages, and an understanding of the cycle may be helpful in identifying these mistakes.

The Metagraph Development Life Cycle (MDLC) is illustrated in the "waterfall" diagram of Figure 10.1. The first stage is *Problem Identification*. This is the identification of an information processing problem and a determination of whether metagraphs provide a reasonable foundation for describing an information processing system for addressing the problem. This consists of identifying (1) the entities of interest, including both the information elements and the assumptions that make up the generating set, (2) the entity aggregates that will make up the invertices and outvertices of the metagraph edges and, (3) the ordered pairs that will define the metagraph edges, including the resources (labels) needed to implement the models or tasks represented by the edges.

The second stage, *Metagraph Construction*, consists of three substages. The first is identification of the generating set. The elements in the generating set will be the data attributes, rule propositions, model varriables, and/or workflow workstations that make up the system being represented. The second substage is the aggregation of these elements into clusters that will become invertices and outvertices. The third substage is the identification of metagraph edges that represent relationships between these clusters and the directions of these relationships.

The third stage, *Requirements Analysis*, consists of two subtasks. The first is structural analysis, which means the specification of information structures – such as metapaths, projections, and contexts – that will be useful in decision making. Of course others can be added as needed, but an understanding of the structures that might be needed may help in determining the information elements (including assumptions and labels) that should be included in the system and the relationships (edges) among these elements. The second subtask is a feasibility study, which usually consists of four components – technical feasibility, economic feasibility, operational feasibility, and schedule feasibility.

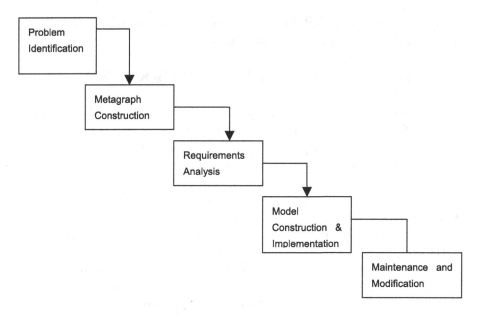

Figure 10.1. The metagraph development life cycle.

The fourth stage, *Model Construction and Implementation*, consists of the construction and implementation of the decision models, stored data relations, business rules, and workflows that make up the system. This is the stage in which much of the "real work" – that is, the data collection, programming, testing, and documentation – takes place. One potential mistake, which should be avoided, is to jump into this stage before the earlier stages, which lay out the ground work for this stage, have been largely completed.

The fifth stage is *Maintenance and Modification.* indexmaintenance and modification Maintenance consists of gradual changes in a decision model, data relation, business rule, or workstation. These may be caused by errors detected in data or software or minor changes in external conditions, such as government relations or the practices of suppliers, customers, or business part-ners. Modification is similar to maintenance except that the changes are more severe. These include such qualitative structural changes as the introduction of new production facilities (and therefore new components of a decision model), major changes in database or rule structures, and new workflow stations.

The maintenance and modification stage is of special importance because much of the cost incurred in the life cycle occurs in this stage. Therefore, the first three stages should be conducted with this stage in mind. The data elements and the edges should be designed for maintainability – for example, element names should be intuitive and procedures and data relations should be well documented. In addition, a metagraph workbench of the type proposed in the next section may be helpful not only in metagraph construction, but also in metagraph maintenance and modification.

2.   TOWARDS A METAGRAPH WORKBENCH

There are three purposes of a DSS analysis workbench. One is to serve as a testbed for analyzing the effectiveness of metagraphs as a representational construct for decision analysis and decision support. It will also require the transformation of analysis and design principles discussed in Section 1 of this chapter into structured procedures and algorithms. Finally, it will enable us to empirically test the viability of any metagraph-based decision process model. While this empirical testing is beyond the scope of this chapter (and this book), it is an important element of any long term research agenda in metagraphs.

The architecture of the system is shown in Figure 10.2. The primary user interface is provided by the *metagraph editor*, which is a graphical user interface for drawing and using metagraphs. The user can create new metagraph edges or recall existing edges from the metagraph store, which maintains all known edges and elements. As each edge is added to a current metagraph, the system updates the corresponding algebraic representation (i.e., the adjacency matrix), and this would be transparent to the user. The current metagraph can be compiled at any time through construction of the transitive closure of the adjacency matrix (i.e., the A^* matrix). Once this is done, the user can obtain additional information such as the presence of any bridges or cycles, and these are displayed in the graphical interface of the editor with color highlighting of the relevant edges. Thus, even during the process of constructing the metagraph, the user can obtain analytical feedback about the DSS resources that are being included in the proposed system. This can help to identify what additional edges (resources) are needed, the potential disadvantages of eliminating any edges (e.g., the elimination of connectivity between certain inputs and outputs), and the multideterminacy conditions caused by cycles.

The system has two internal storage components, a *metagraph store* and an *assumptions store*. The former is used to store all known edges in one or more metagraphs. Edges from one or more metagraphs can be combined into a new metagraph by using the metagraph store (if they are all stored in the same file or in a network of linked files). The metagraph store also enables two or more metagraphs to be combined (using metagraph addition) in the editor. The assumptions store is a propositional database used to maintain assumptions and their valuations. One promising extension is the use of a belief maintenance system as an enhancement to the assumptions store (Raghunathan, Krishnan and May, 1995).

From the metagraph editor the user can access three special modules: the metapath builder, the projection builder, and the context builder.

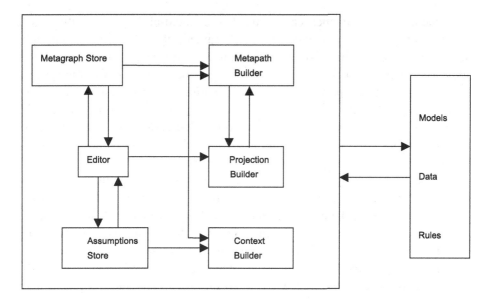

Figure 10.2. Structure of a metagraph-based tool.

- The *metapath builder* constructs all possible metapaths between any two sets of elements. It uses algorithms based on the A^* matrix that is maintained in the metagraph store, once the current metagraph is compiled. The metagraph builder can be accessed several ways. For example, it can be used to construct one or all metapaths between any given element sets specified by the user from the editor. It can also be invoked by the projection builder during the process of constructing projection views of the current metagraph.

- The *projection builder* is used to construct simplified views of the current metagraph over a projection set (of elements) specified by the user. The projection is displayed as a simplified metagraph in a separate window. Since the process of defining a projection involves the identification of a set of dominant metapaths, the projection builder utilizes the metapath builder for this purpose.

- The *context builder* is used to construct simplified views of the current metagraph that identify those edges that are applicable in a specific context (specified by a set of known true and false assumptions). As with a projection, the resulting metagraph can be displayed in a separate window, although an alternative display is by superimposition upon the current metagraph, which emphasizes the fact that the context does not delete any models. The context builder utilizes the A matrix stored in the metagraph store, as well as assumptions maintained in the assumptions store.

A metagraph workbench can be used on a stand-alone basis, with edges being entered manually by the user or maintained internally in the metagraph store. However, a natural extension is to integrate this system with DSS modules such as a model base, data base, and rule base. In this way metagraph construction can be a part of the process of adding models, data, rules, and/or workflows to the DSS. In this case the metagraph representation and associated tools serve as part of the front end to the DSS. An alternative is to view the metagraph-based system as part of the model selection component of an integrated modeling environment, as in (Banerjee and Basu, 1993).

The development of a metagraph workbench would be facilitated by a development in metagraph algorithms not currently addressed in the literature: the identification of planar metagraphs and the representation of planar metagraphs in visually planar form. In graph theory a planar graph is a simple graph or digraph that can be drawn on the plane without intersecting edges, and there are theoretical results for determining whether an arbitrary graph is planar. If it were possible to determine whether an arbitrary metagraph is planar and to render a planar metagraph onto the plane, the visualization of metagraphs, metapaths, projections, and contexts would be facilitated.

3.   METAGRAPHS AND SOCIAL NETWORKS

Social networks are networks of individual people or groups of similarly situated people, typically departments in an organization (Cross, Parker and Sasson, 2003; Cross and Parker, 2004). Social network analysis is a collection of concepts and techniques for representing (usually in the form of simple graphs or digraphs) and analyzing social networks. The resulting concepts concern such topics as kinship, friendship, trust relationships, collaboration, and the sharing and diffusion of information. Of special interest are people and links that are especially prominent – for example, people who have many relationships (e.g., acquaintances, collaborators, etc.) compared with others in the organization and ties that are especially strong. The latter would be represented by a labeled metagraph.

Another important concept, in the context of organizations, is the difference between a hierarchy and a more general network. A hierarchy, often represented by an organization chart, represents the formal structure of an organization, such as reporting relationships and groupings of similarly specialized people and people with similar organizational attributes. People with similar attributes are said to be homophilous, and social and organizational communication often involves homophilous people. A network, on the other hand, represents an informal organizational structure that often crosses organizational stovepipes and is said to describe how the work in an organization really gets done. The resulting simple graph or digraph is sometimes called a sociogram or sociometric diagram.

The application of metagraphs to social networks may be helpful in cross-walking individuals and suborganizations – that is, in treating people both as separate individuals and members of various organizations and suborganizations. These people may communicate directly or through the organizational groups in which they are imbedded. This would require a generalization of metagraphs to symmetric metagraphs, in which the edges are represented as unordered, rather than ordered, pairs of subsets of the generating set. This would allow for undirected links between vertices in the metagraph and would require new definitions and new algorithms concerning metapaths and metagraph connectivity.

Five other areas are as follows:

- The first is cyclic metagraphs. We have briefly described cycles in metagraphs in the context of inference paths in rule-based systems. However, this concept may be even more prominent in social networks. Social and organizational relationships are often cyclic, because they represent iterative processes. An example is budgeting processes in organizations. Hoerver, cyclic metagraphs differ from cyclic simple graphs and digraphs in several respects. For example, the removal of certain edges from a cycle may leave the cycle intact, and algorithms for the identification such edges may be of interest.

- The second is metagraph isomorphism, which is analogous to graph isomorphism. Two metagraphs are isomorphic if they are identical up to a relabeling of their generating sets and edges. Metagraph isomorphism may be useful in identifying two structurally equivalent organizations or other social systems.

- The third is the representation of uncertainty and ambiguity in metagraphs. This would presumably be accomplished by labeling edges and possibly elements with measures of uncertainty and ambiguity. Thus, we may envision stochastic metagraphs and/or fuzzy metagraphs that capture a lack of specificity in knowledge about social actors and their interactions.

- The fourth area is metagraph coloring. Theorems concerning the coloring of simple graphs have been of substantial theoretical interest – for example, in solving the four-color problem concerning planar maps. Analyses of metagraph coloring would be far more complicated, because it would require consideration of what is to be colored – elements and/or edges. However, such a theory and its consequent applications may suggest ways in which disparate people and organizational subgroups may establish boundaries and encourage or resist cooperation.

- The fifth area is the relationship between metapaths and metagraph cuts. In simple graphs a cut is a set of edges whose removal would disconnect two otherwise connected elements or vertices, and the maximum number of edge disjoint paths between two connected vertices is equal to the minimum size of a cut separating the vertices. This has led to some interesting results concerning flows in graphs. It would be interesting to understand any relationships between metapaths and metagraph cuts. This might suggest useful results concerning the flows of materials through metagraph-based logistical structures.

Thus, there are many promising opportunities for new areas of research and application in metagraphs, many possibly not yet anticipated.

4.   AND FINALLY

We believe that the topic of metagraphs is a research goldmine. We have found this topic to be a font of stimulating ideas and mathematical results in several areas of information processing, including decision support (i.e., data management, model management, and rule management) systems and workflow systems. This topic also holds forth the promise additional mathematical results, additional mathematical structures, and additional areas of application – for example, in the modeling of social and organizational systems and possibly other systems as well.

The reason seems to be that we are entering an age of connectivity. People want to communicate and establish systems of centralized, decentralized, and peer-to-peer networks involving people, their organizations, and decision support (data, models, rules) modules. Metagraphs may provide a helpful foundation for these networks.

REFERENCES

Banerjee, S. and Basu, A. (1993). Model type selection in a DSS environment, *Decision Support Systems* **9** (1), 75–89.

Barua, A., Lee, S. C. H. and Whinston, A. B. (1996). The calculus of reengineering, *Information Systems Research* **7** (4), 409–428.

Basu, A. and Blanning, R. W. (1992). Enterprise modeling using metagraphs, In: T. Jelassi, M. R. Klien and W. M. Mayon-White (eds.), *Decision Support Systems: Experiences and Expectations*, North-Holland, pp. 183–199.

Basu, A. and Blanning, R. W. (1992). Metagraphs and Petri nets in model management, In: V. C. Storey and A. B. Whinston (eds.), *Proceedings of the Second Annual Workshop on Information Technologies and Systems (WITS-92)*, Dallas, pp. 64–73.

Basu, A. and Blanning, R. W. (1994). Cycles in metagraphs, In: *Proceedings of the 27th Annual Hawaii International Conference on System Sciences, Vol. III: Information Systems: DSS/Knowledge-Based Systems*, pp. 310–319.

Basu, A. and Blanning, R. W. (1994). Metagraph views in model management, In: *Proceedings of the 4th Annual Workshop on Information Technologies and Systems*, pp. 175–184.

Basu, A. and Blanning, R. W. (1994). Model integration using metagraphs, *Information Systems Research* **5** (3), 195–218.

Basu, A. and Blanning, R. W. (1994). Metagraphs: A tool for modeling decision support systems, *Management Science* **40** (12), 1579–1600.

Basu, A. and Blanning, R. W. (1995). Discovering implicit integrity constraints in rule bases using metagraphs, In: *Proceedings of the 28th Annual Hawaii International Conference on Systems Sciences*, pp. 321–329.

Basu, A. and Blanning, R. W. (1995). Metagraphs, *Omega* **23** (1), 13–25.

Basu, A. and Blanning, R. W. (1996). A metagraph-based DSS analysis workbench, In: *Proceedings of the 29th Annual Hawaii International Conference on Systems Sciences*, pp. 386–395.

Basu, A. and Blanning, R. W. (1997). Metagraph transformations and workflow analysis, In: *Proceedings of the 30th Annual Hawaii International Conference on Systems Science, Vol. IV*, pp. 359–366.

Basu, A. and Blanning, R. W. (1997). Metagraphs in workflow support systems, In: *Proceedings of the 4th Conference of the International Society for Decision Support Systems*, pp. 565–579.

Basu, A. and Blanning, R. W. (1997). A graph-theoretic approach to analyzing knowledge bases containing rules, models and data, *Annals of Operations Research* **75**, 3–23.

Basu, A. and Blanning, R. W. (1998). The analysis of assumptions in model bases using metagraphs, *Management Science* **44** (7), 982–995.

Basu, A. and Blanning, R. W. (1999). Metagraphs in workflow support systems, *Decision Support Systems* **25**, 199–208.

Basu, A. and Blanning, R.W. (2000). A formal approach to workflow analysis, *Information Systems Research* **11** (1), 17–36.

Basu, A. and Blanning, R. W. (2001). Workflow analysis using attributed metagraphs, In: *Proceedings of the 34th Annual Hawaii International Conference on Systems Science* (CD/ROM, 9 pages).

Basu, A. and Blanning, R. W. (2003). Synthesis and decomposition of processes in organizations, *Information Systems Research* **14** (4), 337–355.

Basu, A., Blanning, R. W. and Shtub, A. (1997). Metagraphs in hierarchical modeling, *Management Science* **43** (5), 623–639.

Berge, C. (1985). *Graphs*, 2nd edn., North-Holland, Amsterdam.

Berge, C. (1989). *Hypergraphs*, North-Holland, Amsterdam.

Blanning, R. W. (2003). Model building process, In: H. Bidgoli, S. B. Eom and A. Prestage (eds.), *Encyclopedia of Information Systems, Vol. 3*, Elsevier Science, Amsterdam, pp. 181–191.

Cross, R. and Parker, A. (2004). *The Hidden Power of Social Networks*, Harvard Business School Press, Boston.

Cross, R., Parker, A. and Sasson, L. (eds.) (2003). *Networks in the Knowledge Economy*, Oxford University Press, Oxford.

Curtis, B., Kellner, M. I. and Over, J. (1992). Process modeling, *Communications of the ACM* **35** (9), 75–90.

Date, C. J. (1995). *An Introduction to Database Systems*, 6th edn., Addison-Wesley, Reading, MA.

Davenport, T. H. (1993). *Process Innovation: Reengineering Work through Information Technology*, Harvard Business School Press, Boston.

de Kleer, J. (1986). An assumption-based TMS, *Artificial Intelligence* **28** (2), 127–162.

Dennis, A. and Wixom, B. H. (2000). *Systems Analysis and Design*, Wiley, New York.

Gallo, G., Longo, G., Pallottino, S. and Nguyen, S. (1993). Directed hypergraphs and applications, *Discrete Applied Mathematics* **42** (2–3), 177–201.

Grant, J. and Minker, J. (1991). Integrity constraints in knowledge Basu systems, Technical Report CS-TR-2223, University of Maryland, College Park, MD.

Hammer, M. (1996). *Beyond Reengineering*, Harper Business, New York.

Hammer, M. and Champy, J. (1993). *Reengineering the Corporation: A Manifesto for Business Revolution*, Harper Business, New York.

Harel, D. (1987). Statecharts: A visual formalism for complex systems, *Science of Computer Programming* **8** (3), 231–274.

Harel, D. (1988). On visual formalisms, *Communications of the ACM* **31** (5), 514–530.

Hoffer, J. A., George, J. F. and Valacich, J. S. (1999). *Modern Systems Analysis and Design*, Addison-Wesley, Reading.

Illarramendi, A., Blanco, J. M. and Goñi, A. (1994). Making knowledge base systems more efficient: A method to detect inconsistent queries, *IEEE Trans. on Knowledge and Data Engineering* **6** (4), 634–639.

Khoshafian, S. and Buckiewicz, M. (1995). *Introduction to Groupware, Workflow, and Workgroup Computing*, Wiley, New York, Chapter 5, pp. 207–258.

Kwan, M. and Balasubramanian, P. R. (1997). Dynamic workflow management: A framework for modeling workflows, In: *Proceedings of the 30th Hawaii International Conference on System Sciences, Vol. IV, Maui, HI*, pp. 367–376.

Marschak, R. T. (1995). Workflow: Applying automation to group processes, In: D. Coleman and R. Khanna (eds.), *Groupware: Technologies and Applications*, Prentice-Hall, Upper Saddle River, Chapter 3, pp. 71–77.

Moder, J. J., Phillips, C. R. and Davis, E. N. (1983). *Project Management with CPM, PERT, and Precedence Diagramming*, Van Nostrand Reinhold, New York.

Peterson, J. L. (1981). *Petri Net Theory and the Modeling of Systems*, Prentice-Hall, Englewood Cliffs.

Raghunathan, S., Krishnan, R. and May, J. H. (1995). On using belief maintanance systems to assist mathematical modeling, *IEEE Transactions on Systems, Man and Cybernetics* **26** (2), 287–303.

Ramaswami, M., Sarkar, S. and Chen, Y.-S. (1997). Using directed hypergraphs to verify rule-based expert systems, *IEEE Transactions on Knowledge and Data Engineering* **9** (2), 221–237.

Thomson, V. J. (1995). Process monitoring for continuous improvement, In: J. Browne and D. O'Sullivan (eds.), *Reengineering the Enterprise*, Chapman and Hall, London.

INDEX OF DEFINITIONS

° operator	Definition 2.11
⊗ operator	Definition 4.8
Adjacency matrix	Definition 2.6
Bridge	Definition 3.6
Cat (catenation operator)	Definition 2.9
Coinput	Definition 2.4
Components of ordered triple	Definition 2.8
Composition	Definition 4.2
Conditional metagraph	Definition 5.1
Conditional metapath	Definition 5.3
Connected	Definition 5.6
Context metagraph	Definition 5.2
Cooutput	Definition 2.4
Correspondance	Definition 8.2
Critical edge	Definition 9.3
Critical invertex	Definition 9.1
Cutset	Definition 3.5
Cyclic element within a metapath	Definition 8.3
Dominance	Definition 3.3
Edge	Definition 2.2
Edge-dominant	Definition 3.1
Element flow metagraph	Definition 4.7
Element used in a metapath	Definition 8.4
Flow composition	Definition 4.7
Flow content	Definition 4.7
Fully connected	Definition 5.6
Fully and redundantly connected	Definition 5.6
Generating set	Definition 2.1
Incidence matrix	Definition 2.13
Independence	Definition 6.1
Input-dominant	Definitions 2.14 and 3.2
Input independence	Definition 6.1
Interpretation	Definition 5.5
Inverse metagraph	Definition 4.6

Invertex	Definition 2.2
Length	Definition 2.4
Metagraph	Definition 2.3
Metagraph dominance	Definition 4.4
Metagraph equivalence	Definition 4.5
Metapath	Definition 2.5
Metapath dominance	Definition 4.3
Mutual independence	Definition 6.2
Non-redundant conditional metagraph	Definition 5.7
Non-redundant edge	Definition 3.4
Non-reduntantly connected	Definition 5.6
Output independence	Definition 6.1
Outvertex	Definition 2.2
Product (of two adjacency matrices)	Definition 2.12
Projection of a conditional metagraph	Definition 4.1
Projection of a metagraph	Definition 5.4
Rule base	Definition 8.1
Set of input propositions	Definition 5.3
Set of intermediate propositions	Definition 5.3
Set of relevant propositions	Definition 5.3
Simple path	Definition 2.4
Slack	Definition 9.2
Source (of a conditional metagraph)	Definition 5.3
Sub-metagraph	Definition 6.1
Sum (of adjacency matrices)	Definition 2.7
Target (of a conditional metagraph)	Definition 5.3
Trnc (truncation operator)	Definition 2.10
Used (in a metapath)	Definition 8.4

Index

accounting model 88, 108
– simple 108
activity duration 55
acyclic metagraph 71, 99, 101, 103, 138
acyclic metapath 12, 100, 102, 104, 107, 115
acyclic path 103
acyclic proposition 108
adjacency matrix 17–21, 23, 28, 31, 36, 37, 54, 72–74, 92, 94, 104, 109, 110, 114, 115, 128, 129, 131, 135, 156
– closure of 21, 22, 73, 74, 156
– multiplication of 19, 21
– set of 23
algebraic properties 99
allocation problem 133
assumption 72, 86, 89, 93–95, 122, 125, 153
– initial 87
– intermediate 95
– set of 87, 94
assumption-based analysis 95
assumptions in model bases 72
assumptions store 156
atomic models 75
attribute
– qualitative 53, 54
– quantitative 55
attributed metagraph 10, 53, 134, 146

behavioral modeling 120
Boolean information elements 119
Boolean truth value 105
Boolean variable 98
bridge 27, 29–31, 71, 72, 76, 130, 156

capacity to produce 74
closure 21, 22, 36, 104, 115, 135
closure matrix 110, 133, 135, 138
coinputs 15
combined metagraph 20
combining model bases 79
commutativity property 93
composite model 11
composition 35, 36, 38, 49, 78–80

concept of independence 144
conditional metagraph 11, 55–62, 72, 86, 90–95, 145, 153
– features of 92
conditional metapath 57, 58, 87–89, 94
connected 62
– fully 62
– non-redundantly 62
connectivity 61, 62, 72
connectivity properties 10, 11, 16, 27, 99, 130
conservative approach 24
constrained model 105
constraint 86
context 56, 58, 59, 72, 90, 91, 93, 94, 130, 153, 154
context builder 157
context metagraph 56, 57, 61, 62, 130
cooutputs 15
cost estimating metagraph 81
cost estimating relationships 45, 81, 85
cost estimation 152
cost model 72
critical assumption 89
critical edge 151, 152
critical element 150–152
critical invertex 150
critical path 151, 152
– calculations 118
critical task 151, 152
critical vertices 151
cutset 29
cycle 27, 109, 130, 156
cyclic digraph 159
cyclic metagraph 71, 74, 75, 102, 159
cyclic metapath 103, 104, 107
cyclic model 74
cyclic proposition 108
cyclic simple graph 159

data attributes 12, 154
data base 12, 97, 105
data flow diagram 1, 3
data management 11, 12, 160

– systems 9, 91
data relations 98, 105, 115
– with domain constraints 105
– with integrity constraint 97
decision model 12, 98, 115
decision support 160
– system 1, 9, 12, 98, 153
decomposable organization 144
decomposition
– of processes 143
– of workflows 12
demand model 74
diagramming conventions 1
digraph *see also* directed graph 158, 159
– cyclic 159
directed graph 1, 4–7, 10, 16
directed hypergraph 7, 9
disjoint sets 30
dominance 27, 40–43, 45
– mutual 44
dominant metapath 27, 28, 34–36, 38, 40,
    59, 76, 85, 93, 129
dual metagraph 46

edge 15, 153
– non-redundant 28, 29
– redundant 28
edge representation 53
edge-centered representation 46
edge-disjoint metapath 29–31
edge-disjoint sub-metagraphs 66
edge-dominance 27
edge-dominant metapath 27–30
element flow metagraph 33, 48, 51, 134
entity-relationship approach 2
entity-relationship diagram 1
equality predicate 105, 106, 108
equivalence 44, 84
expected service life 81
explicit integrity constraint 112

false proposition
– set of 62
feasibility study 154
financial model 72, 88
flow composition 49, 51
forecasting model 88
fully connected 62
functional dependency 2, 106
– diagram 1, 2

functional interaction 133
functional modeling 120, 121

generating set 12, 15, 18, 19, 21, 36, 43, 44,
    46, 72, 76, 83, 86, 98, 100, 123–125,
    131, 153, 154
– overlapping 71, 79
– union of 71
graph
– directed 4–7, 10, 16
– simple 4, 5, 10
graph isomorphism 159
graph structure 4, 5
gross national product 74

hidden intermediate variable 35
hierarchical modeling 11, 71, 76
higraph *see also* hierarchical graph 1, 4
Horn clause 12
– logic 100
– rules 99
hypergraph 1, 4, 6
– directed 7, 9

implicit integrity constraint 12, 111–113
incidence matrix 23, 24, 46, 47, 51, 64, 127
independence 65, 67
– concept of 144
– input 65
– output 65
independence conditions 145
independence of metagraph 145
independent metagraph 65
independent metapaths
– set of 67
independent sub-metagraphs 65, 68
independent sub-processes 143
independent workflows 143, 144
inflation rate 72
information
– element 122–125, 130, 136
– – Boolean 119
– – time-critical 148
– flow 119
– interactions 128
– loss of 40, 42
– processing systems 71, 117
– structures 97
informational modeling 120, 121
initial assumptions 87

– set of 87
input independence 65
input propositions
– set of 58
input variables 4
input dominance 27
input-dominant metapath 24, 27–29
input-independent metagraph 65
input-to-output mapping 71
integrity constraint 110–114
– explicit 112
– implicit 12, 111–113
– potential 111
– set of 114
inter-task scheduling 151
interactions among components 123
intermediate assumptions 87, 95
– set of 87
intermediate propositions
– set of 58
intermediate variables 4
– hidden 35
interpretation 61
intra-task scheduling 151
inverse 46
inverse metagraph 46–48, 131
invertex 15
– critical 150

joint metagraph 41

knowledge base 105–110
knowledge-based systems 111

labeled metagraph 158
life cycle 153, 155
– concept 154
– costing 76–78, 80, 82
– – calculations 45, 85
– metagraph development 154, 155
– model development 153
– of metagraph construction 12
– of metagraph implementation 12
– system development 153
loss of information 40, 42

maintenance and modification 155
management of data bases 71
management of decision models 71
management of workflow systems 71

many-to-many mapping 133
marketing model 108
matrix organization 144
metagraph 8, 9, 15, 16
– acyclic 71, 99, 101, 103, 138
– algebraic structure 15
– attributed 53, 134, 146
– basic properties 15
– combined 20
– conditional 11, 55–62, 72, 86, 90–95, 145, 153
– – features of 92
– – non-redundant 62
– context 56, 57, 61, 62, 130
– cost estimating 81, 85
– cyclic 71, 74, 75, 102, 159
– dual 46
– element flow 33, 48–51, 134
– independence of 145
– independent 65
– input-independent 65
– inverse 46–48, 131
– joint 41
– labeled 158
– non-redundant 64, 142
– output-independent 65
– process 145
– pseudo-dual of 33
– resource independence 118
– resource interaction 118, 134, 135, 146
– simple 62
– symmetric 159
– task interaction 118, 131, 132
– time-constrained workflow 148
– workflow 130, 152
metagraph applications 1
metagraph coloring 159
metagraph connectivity 153
metagraph construction 154
metagraph development life cycle 154, 155
metagraph editor 156
metagraph isomorphism 159
metagraph maintenance and modification 156
metagraph modeling 12, 153
metagraph representation 23, 106–110, 130, 138, 142
metagraph store 156
metagraph transformations 10, 153
metagraph workbench 12, 153, 156, 158

metagraph-based analysis   105, 148
metagraph-based decision   156
metagraph-based DSS   110
metagraph-based logistical structures   160
metagraph-based tool   157
metagraphs
– multiple   39
– mutually independent   66
– related   40
metapath   16, 154
– acyclic   12, 100, 102, 104, 107, 115
– conditional   57, 58, 87–89, 94
– cyclic   103, 104, 107
– dominant   27, 28, 34–36, 38, 40, 59, 76,
      85, 93, 129
– edge-disjoint   29–31
– edge-dominant   27–30
– input-dominant   24, 25, 27–29
– source-dominant   129
metapath builder   157
metapath source variables   95
metapath target variables   95
metapaths
– independent
– – set of   67
– multiple   71, 76
model
– constrained   105
– unconstrained   105
model base   105
model construction and implementation   155
model development life cycle   153
model management   11, 71, 93, 160
– systems   91
model selection and integration   71, 74
model varriables   154
models   95
models as metagraphs   72
modular workgroups   144, 145
monolithic organization   144, 145
multiple cutsets   29
multiple metagraphs   39
multiple metapaths   71, 76
multiple simple paths   76
multiplication of adjacency matrices   21
mutual dominance   44, 84
mutually independent metagraphs   66

net income   72
non-critical element   150

non-redundancy   142
non-redundant edge   28, 29
non-redundant metagraph   64, 142
non-redundantly connected   62
non-zero slack   151

ordered triple   19
organizational modeling   120, 121
output independence   65
output variables   4
output-independent metagraph   65
outvertex   15
overall economic activity   74
overlapping generating sets   71, 79

path
– acyclic   103
– simple   16, 17
Petri net   1, 4, 7, 8
potential integrity constraint   111
price   74
price–volume relationship   86
pricing model   108
problem identification   154
procedure
– Check-Independence   67, 68
– Critical-Path   149, 151
– EFM   50, 134
– Inverse   47, 131
– PIPO   68
– Projbuild   37–39
– Proof   100
process   122
– analysis   11, 12
– decomposition   145
– metagraph   145
project management systems   1
projection   33–40, 44, 58–60, 72, 76–78,
      80–83, 92, 93, 130, 153, 154
projection builder   157
projection of union   72
projections
– union of   72
properties of sub-metagraphs   153
proposition   12, 56, 57, 59, 86, 91, 98, 99,
      105, 130
– acyclic   108
– cyclic   108
– false

– – set of   62
– input
– – set of   58
– intermediate
– – set of   58
– relevant   58
– – set of   58
– set of   55, 99, 107, 110
– true
– – set of   62
proposition definition   105
proposition set   63
pseudo-dual of metagraph   33
pure inputs   63, 64
pure outputs   63, 64

qualitative attributes   53, 54
quantitative attributes   55

redundancy   61, 143
redundant edge   28
related metagraphs   40
relationships among concepts   3
relationships among items   3
relationships among multiple elements   6
relationships among variables   5
relationships between sets of variables   7
relevant propositions   58
– set of   58
report   122
representation
– edge-centered   46
representation of decision models   71
requirements analysis   154
resource   122, 123, 125, 133, 136, 152, 156
resource independence metagraph   118
resource interaction   133, 145
– metagraph   118, 134, 135, 146
resulting sales volume   74
rule   105
rule base   12, 97, 99, 100, 105, 107, 111,
   112, 114
rule base as metagraph   99
rule management   11, 12, 97, 160
rule propositions   154
rule-based inference   107
rule-based systems   9, 159
rules   115
– Horn clause   99
– set of   99

semantic equivalence   53
semantic net   1, 4
semantics   106
sensitivity analysis   95
sequence of edges   16
service life estimation   152
set of assumptions   94
set of edges   15, 16
set of false propositions   62
set of independent metapaths   67
set of input propositions   58
set of integrity constraints   114
set of intermediate propositions   58
set of propositions   55, 63, 99, 107, 110
set of relevant propositions   58
set of rules   99
set of true propositions   62
set of variables   55
set-to-set mapping   71, 153
simple accounting model   108
simple graph   1, 4, 5, 10, 158, 159
– cyclic   159
– coloring of   159
simple metagraph   62
simple path   15–17
simple paths
– multiple   76
single model base   79
singleton cutset   29
singleton set   17
slack   151
social network   153, 158
– modeling   12
sociometric diagram   159
source   122, 124, 129
source-dominant metapath   129
structural analysis   154
structure of databases   2
sub-metagraph
– properties of   153
sub-metagraph independence   143
sub-metagraphs   11
– edge-disjoint   66
– independent   65, 68
– independence of   143, 144
sub-processes
– independent   143
supply model   74

symmetric metagraph   159
synthesis of processes   137
synthesis of workflows   12
systems development life cycle   153

target   122, 124, 129
task   119, 122, 123, 125, 136
– executed   122
task duration   55
task interaction   131
– metagraph   118, 131, 132
temporal constraints   133
time-constrained workflow metagraph   148
time-critical information elements   148
time-critical workflows   118, 146
transaction processing systems   1
transactional modeling   120
true propositions
– set of   62
truth value   105
– Boolean   105
types of relationship   98

unconstrained model   105, 110
union of projections   72

value   75
variables   56, 77, 86, 105
– set of   55
vertex representation   53, 54

well-structured process   142
workflow   122
– analysis   11, 12
– duration   55
– management systems   9
– metagraph   130, 152
– metapath
– – dominant   133
– system   1, 117, 122, 153, 160
– time-critical   118
– workstations   154
workflows
– independent   143

zero slack   151